Springer Theses

Recognizing Outstanding Ph.D. Research

For further volumes:
http://www.springer.com/series/8790

Aims and Scope

The series "Springer Theses" brings together a selection of the very best Ph.D. theses from around the world and across the physical sciences. Nominated and endorsed by two recognized specialists, each published volume has been selected for its scientific excellence and the high impact of its contents for the pertinent field of research. For greater accessibility to non-specialists, the published versions include an extended introduction, as well as a foreword by the student's supervisor explaining the special relevance of the work for the field. As a whole, the series will provide a valuable resource both for newcomers to the research fields described, and for other scientists seeking detailed background information on special questions. Finally, it provides an accredited documentation of the valuable contributions made by today's younger generation of scientists.

Theses are accepted into the series by invited nomination only and must fulfill all of the following criteria

- They must be written in good English.
- The topic of should fall within the confines of Chemistry, Physics and related interdisciplinary fields such as Materials, Nanoscience, Chemical Engineering, Complex Systems and Biophysics.
- The work reported in the thesis must represent a significant scientific advance.
- If the thesis includes previously published material, permission to reproduce this must be gained from the respective copyright holder.
- They must have been examined and passed during the 12 months prior to nomination.
- Each thesis should include a foreword by the supervisor outlining the significance of its content.
- The theses should have a clearly defined structure including and introduction accessible to scientists not expert in that particular field.

Jonathan M. Taylor

Optical Binding Phenomena: Observations and Mechanisms

Doctoral Thesis Accepted by Durham University, UK

 Springer

Author
Dr. Jonathan M. Taylor
Department of Physics
Centre for Advanced Instrumentation
University of Durham
South Road
Durham DH1 3LE
UK
e-mail: j.m.taylor@durham.ac.uk

Supervisor
Dr. Gordon D. Love
Department of Physics
Centre for Advanced Instrumentation
University of Durham
South Road
Durham DH1 3LE
UK
e-mail: g.d.love@durham.ac.uk

ISSN 2190-5053
ISBN 978-3-642-21194-2
DOI 10.1007/978-3-642-21195-9
Springer Heidelberg Dordrecht London New York

e-ISSN 2190-5061
ISBN 978-3-642-21195-9 (eBook)

Cover design: eStudio Calamar, Berlin/Figueres

Printed on acid-free paper

Springer is part of Springer Science+Business Media (www.springer.com)

Supervisor's Foreword

Optical binding is a remarkable phenomenon whereby the presence of a micro-particle in an optical landscape can influence the light distribution which in turn can affect the position of the particles. As a simple example two laser beams pointing towards each other a counter propagating trap form a potential well analogous to springs attached to particles. If one increases the number of particles in the trap then naively one would expect all the particles to collect in the centre of the well. However, the effect of optical binding means that the presence of one particle affects the distribution of light with the effect that the particles can arrange themselves into arrays as well as displaying other interesting phenomena. Optical binding is both of theoretical interest and has applications in micromanipulation and assembly.

The work described in this Thesis combines experimental results with a sophisticated generalized Lorenz–Mie scattering model implementation developed by Jonathan Taylor as part of his Ph.D., the capabilities and speed of which make it possible to study time evolution of a multiple particle system. The work begins with an introduction to generalized Lorentz–Mie scattering theory written in a manner accessible to the non-specialist, and including some novel results. It then compares experimental and theoretical results for a range of trap configurations, not just demonstrating agreement between theory and experiment, but using the model to provide insight and interpretation of the experimental results.

This thesis therefore contains results of the full spectrum of physics activities: rigorous electromagnetic theory, advanced computer simulations, extremely sensitive experimental results, and physical analysis and interpretation of those results. For example, he has shown that particles can undergo circulatory motion in a trap where all the parameters are constant. This remarkable result was described both by his computer model, and by a simple heuristic model showing an intuitive explanation, as well as being observed experimentally.

Durham, March 2011

Dr. Gordon D. Love

Acknowledgments

Most of all I would like to thank my supervisor Gordon Love for his support and guidance throughout the four years work leading to this thesis, for having confidence in the unexpected tangents the project took...and indeed for persuading me to work on optical trapping in the first place, rather than on his "bait" Ph.D. project!

I am also very grateful to the various people I have been fortunate enough to collaborate with or correspond with, including Colin Bain, Pavel Zemánek, Jack Ng and Reuven Gordon.

Finally, I must acknowledge the "help" of the various inhabitants of room 125b, a room which has seen some great mysteries as well as great engineering achievements, and the invaluable assistance of MH who had a beer ready the one time when I most needed it.

Contents

Chapter 1
Introduction

1.1 Motivation

This thesis investigates phenomena occurring when multiple particles are confined in the same optical trap, leading to light-mediated interactions between the trapped particles (*optical binding*). These interactions are not only of interest in terms of the fundamental optical physics involved, but also have many practical implications for micro-manipulation of dielectric particles. Multiple particles may be manipulated for the purpose of microstructure construction, using either:

- individual optical tweezers, where any inter-particle interactions are an undesirable side-effect [1]; or
- broader optical fields, where inter-particle interactions can be critical to the self-assembly of a structure [2–4]

Multiple particle interactions are also of interest in optical lattices, optical sorting and optical transport [5–8].

The aim of this thesis is to develop a better understanding of the physical mechanisms underlying the optical binding interaction. The intention, where possible, is to discuss the phenomena in a physically intuitive manner, drawing insight from a rigourous analysis of the system to develop simple, easily-understood explanations for the effects observed.

This thesis presents results on the optical binding of optically-trapped microparticles. A sophisticated Mie scattering model is developed, capable of performing time-evolution simulations of a multi-particle system. This is used to analyse and interpret experimental results in evanescent and Gaussian beam traps, and to develop simple, intuitive explanations for the observed phenomena. Novel trapped states are reported, that do not conform to the symmetry of the underlying trap. A common theme throughout this thesis is the "emergent" phenomena that occur when multiple particles are trapped together, which cannot easily be predicted by considering each particle in isolation.

J. M. Taylor, *Optical Binding Phenomena: Observations and Mechanisms*,
Springer Theses, DOI: 10.1007/978-3-642-21195-9_1,
© Springer-Verlag Berlin Heidelberg 2011

1.1.1 Experimental History of Optical Trapping and Binding

Optical confinement of micron-sized particles in two dimensions using a focused laser beam was first demonstrated by Ashkin et al. in 1970 [9], and subsequently extended to three-dimensional confinement using two counter-propagating beams, originally in the form of an ion trap [10]. A significant development from this has been the field of *optical tweezers*, where a high numerical aperture laser beam is used to trap and manipulate a microparticle. A relatively independent development has been the investigation of interactions between multiple trapped particles (*optical binding*, generally in low numerical aperture configurations).

The term *optical binding* was first introduced in [11], referring to interference effects between the light scattered by a single trapped particle and the background laser trapping light. This interference strongly modifies the electromagnetic field around the particle, and the field experienced by a second nearby particle (and hence the force on it) is different to that which either of the individual particles would experience in isolation. The presence of interference fringes produced by the scattered light tends to cause particles to be trapped and "bound" at discrete inter-particle distances that, depending on the trap geometry, will often be multiples of the laser wavelength (lateral geometries) or half the laser wavelength (longitudinal geometries).

"Optical binding" has more generally been used to refer to any experimental phenomenon whereby multiple trapped particles interact to form well-defined, reproducible, bound structures. We will see in Chap. 4, however, that in some cases the phenomena are not in fact caused by binding in a strict interference-based definition, as was the case in [11].

This thesis is almost exclusively concerned with optical binding effects, rather than optical tweezers, but we will return to this distinction in Chap. 5 where we will discuss the implications of optical binding for high numerical aperture optical tweezers experiments. Consequently the brief review of the field given in the following sections will focus on optical binding, and areas of optical tweezing where inter-particle interactions are expected to be relevant. In addition to this, subsequent chapters give a further introduction to the literature relevant to the chapter topic.

1.1.1.1 Optical Binding

Optical binding between multiple optically trapped particles was reported by Burns et al. in 1989 in a lateral configuration where the particles lay in a plane perpendicular to the direction of propagation of the trapping laser beam [11, 12]. In this case the confinement was only two-dimensional. Interestingly, this lateral configuration has never to our knowledge been extended to three dimensions by

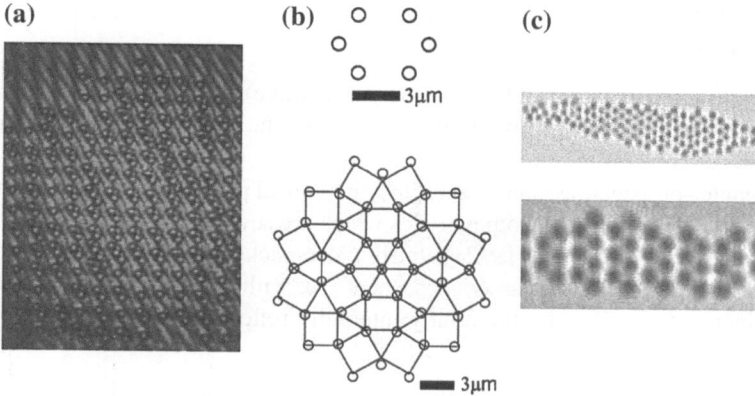

Fig. 1.1 Examples of optically bound particle clusters (**a**), two-dimensional arrays in [12] (**b**), relatively widely-spaced "molecules" in [13] (**c**), fairly close-packed regular structures in an evanescent wave trap, in [20]

the use of two counter-propagating beams[1], although such a configuration has been studied theoretically [13].

It was not until 2002 that two groups [14, 15] independently reported longitudinal interactions between multiple particles in a counter-propagating Gaussian beam trap (where the particles lie parallel to the direction of propagation of the trapping beams). This is the type of configuration that we will mostly focus on in this thesis.

Optical binding has been observed in theory and experiment to give rise to a rich tapestry of nonlinear static and dynamic behaviour, including:

- chain formation [14–16], where trapped particles in a counter-propagating Gaussian beam trap form linear chains, with the spacing between the particles dictated by the optical binding interaction.
- bistability [17, 18], where for some experimental parameters there are multiple stable spacings between a given number of particles. For a Gaussian beam trap this generally means one "standard" spacing, which can be explained using our model in Chap. 4, and one very close spacing where the particles are almost touching, where the repulsion is produced by complicated near-field effects.
- two-dimensional "crystal" arrays [12, 13, 19–21], where sub-micron-sized particles form two-dimensional regular structures perpendicular to the direction

---

[1] This may be because there are many interference fringes formed, with each of these defining a different plane of trapping perpendicular to the beam axes. In order to form optically bound clusters in a single plane there must be some way of ensuring all the particles are located in the same plane. This would probably require manual loading using "helper" optical tweezers.

of propagation of the trapping beam(s). These have been observed in a number of different situations (see Fig. 1.1):

- broad Gaussian beam [12] (pseudo-plane wave): fairly close-packed regular structures trapped in two-dimensions, with the third dimension of confinement provided by the walls of the cell;
- counter-propagating plane waves[13]: theoretical prediction of two-dimensional "molecules" formed from particles which are several diameters apart;
- evanescent wave trap [19–21]: fairly close-packed regular structures trapped in two dimensions in an evanescent wave, with the third dimension of confinement provided by the totally internally reflecting substrate.

As we will discuss in Chap. 3 these scenarios are particularly challenging to understand, not only because of the large numbers of interacting particles but also because of the combination of optical binding, physical close-packing constraints and possible electrostatic charges, all of which can have a significant influence on the structures formed.

- periodic particle motion and instabilities [13, 22–25] where, despite the overdamped nature of the system, driven harmonic oscillations and instabilities can be observed.

Possible future areas of development for optical binding include:

- aerosol trapping: recent preliminary results have reported optical trapping and binding of aerosol particles in air [26–30].
- large space-based structures: it has been proposed that the self-organising properties of optically bound particles could be use to self-assemble enormous planar structures in space, which could be used as telescope mirrors or solar sails [3]. Although this would be a very exciting and spectacular application for optical tweezing and binding, our findings presented here cast doubt on the scalability of this proposed technique. We will return to this question in Chap. 5.

1.1.1.2 Optical Tweezers

Confinement of a particle using a single-beam high numerical aperture optical tweezers set up was demonstrated in 1986 [31]. Since then optical tweezing and micromanipulation has developed into a broad field with many applications. Comprehensive reviews of micromanipulation can be found in [7, 32].

In the context of the phenomena discussed in this thesis, our main interest in optical tweezing is where multiple particles are being trapped together. Examples include:

- Time-sharing to produce multiple optical tweezers from a single laser beam [33]. This is a simple system to design, but suffers from scalability problems, since as more trapping sites are added the fraction of time that the laser is assigned to each trapping site is reduced, and the number of trapped particles is limited by the field of view of the trapping objective lens. This reduces the strength of the trap and increases the tendency for particles to diffuse out of the trap. It does however have the advantage that, since only one particle is illuminated at a time, optical binding effects should be negligible.

- Holographic optical tweezers which can generate arbitrary "optical landscapes" [5, 6]. Particles will interact with the background optical field and can be trapped. If the particles are close enough together, optical binding interactions may be extremely important in such a setup. An example of a simple optical landscape is the interference fringes discussed in Chap. 3, for which we show that optical binding effects are absolutely critical to understanding the trapped structures which form.
- Measurement of hydrodynamic interactions between multiple trapped particles [34, 35]. This is of dual interest in the context of this thesis: an accurate description of hydrodynamic interactions may be important for a detailed understanding of close-packed trapped structures (see Chap. 3), and when performing hydrodynamic measurements it is important to understand whether optical binding effects may be present because, if so, this needs to be taken into account in the analysis of the experiment.
- Bessel beam guiding. The non-diffracting nature of Bessel beams make them useful for trapping and transport of multiple particles [36], and also enables easy trapping and transport of multiple particles [37, 38].
- Extended trapping regions such as those generated by vortex-like traps [39], where there has been little attention paid to any optical binding interactions that may be present between multiple particles.

In many such optical tweezers arrangements, optical binding is not an intended consequence of the setup. Rather, it is seen as an inconvenience—if indeed it is considered at all. As the numbers of trapped particles (particularly particles with higher radii and refractive indices) increases, the potential for significant optical binding effects will increase.

1.1.2 Interpreting Binding Experiments

Although there have been intermittent attempts to theoretically model the optical binding behaviour between multiple trapped particles, this has proved extremely challenging. One of the main challenges is knowing the experimental parameters well enough to carry out a simulation which can reasonably be expected to agree with experiments.

McGloin et al. developed a simple model based on consideration of the scattering force produced by the light "refocused" by successive particles in a chain [18, 40, 41]. Although we have revisited and extended such a model in Chap. 4, there do appear to be shortcomings with the comparison of experiment and theory presented in [40]. Reasonable agreement was presented between their experimental results and their theoretical model, but in contrast Mie scattering theory predicts significantly different results for the experimental parameters quoted in that paper (different particle spacings, or even whether stable chains are supported or not). This highlights the sensitivity of optical binding experiments to the exact experimental parameters, some of which are often hard to directly measure.

Direct visualization of the modification to the background laser field was demonstrated in [16], using two-photon fluorescence to directly image the intensity of the electromagnetic field around chains of trapped particles. Here good agreement was shown between a paraxial field propagation model and the experimental results. Unfortunately a lack of detail in some of the parameters used has meant we have been unable to make a direct comparison between their model and our Mie scattering model.

It was only in 2008 that very good agreement was demonstrated between Mie scattering theory and experimental results from fibre-based optical trapping experiments at the University of Victoria, Canada (Gordon et al. [24, 42]). It is interesting to note that this significant milestone was achieved using a fibre-based trap for the experimental measurements: this considerably reduces the challenges of alignment and beam quality. It is these factors which prove a considerable obstacle to accurate quantitative measurements in lens-based counter-propagating beam traps.

In addition to this, a model based on coupled dipole calculations has also recently been used to inform a simple explanation of the mechanisms that led to chain formation in a counter-propagating Bessel beam trap [37]. It is worth noting that this mechanism is very different to that which we discuss for Gaussian beam traps in Chap. 4, despite the apparent similarity in the trap configuration and nature of the particle chains formed.

The results of Gordon et al. have validated the use of Mie scattering theory as the gold standard in the modelling of optical trapping and binding, and that achievement in [42] provides important support for our choice of techniques used in this thesis, as well as providing an independent comparison for our Mie scattering computer code, developed separately to theirs. The motivation behind this thesis is echoed by Gordon et al.:

> Currently, no theory has explained fully the occurrence of inhomogeneous particle spacing, both for a particle number dependency and a dependence on inter-array particle positions (i.e. inner and outer inter-particle spacings differ for a fixed number particle array), and the spontaneous onset of oscillations observed in the dual beam trap [24].

Chapter 4 of this thesis will address all these questions.

1.2 Synopsis

The aim of this thesis is to offer physical interpretations and insights into optical binding phenomena observed in optical traps. The common theme throughout is the concept of interactions between multiple particles, and the fact that they give rise to behaviours which are substantially different from the behaviours of isolated trapped particles. The observations discussed here cannot be predicted simply by studying isolated particles.

Some of the experimental results which form the starting point for the investigation are from our own experiments. The experimental results referred to in Chap. 3 are the work of our collaborators (whose contribution is made clear in the text where this is the case). We analyse and interpret these results using Mie scattering theory. We developed the computer model used in order to have code which meets the needs of the analysis. In addition to Mie scattering simulations, the results in Chap. 4 are interpreted using a simpler conceptual model which offers more insights into the physical mechanisms underlying the effects.

Chapter 2 describes Mie scattering theory, and briefly discusses our computer model. Some of the content of this chapter draws together results published by a variety of authors into a coherent whole, and some of the results and approaches within the chapter are entirely novel work. The distinction between these is made in the introduction to the chapter, which also briefly summarizes the history of Mie scattering theory and its generalizations.

Chapter 3 discusses evanescent wave trapping experiments, and the two-dimensional "crystal structures" of nanoparticles reported by Bain et al. [19, 20] which are formed by optical binding effects. We explain some of the structures through Mie scattering theory, and discuss the relative contributions of optically-induced forces and collisional interactions on the nature of the structures which are formed.

Chapter 4 considers a trap formed by two counter-propagating Gaussian beams. We describe new experimental results showing trapping configurations which do not conform to the underlying symmetry of the trap, resulting in both stationary modes and non-stationary trapping modes in which particles circulate around the trap away from the common beam axis. We show that such configurations are predicted by Mie scattering theory, along with the more familiar simpler on-axis stationary trapped chains. We discuss the physical interactions which give rise to each of these configurations, and show that the stationary chains can be fairly well described using an extremely simple conceptual scalar model of the light-mediated interaction.

Finally, Chap. 5 summarizes the conclusions drawn from this work, and topics worthy of further investigation.

References

1. Castelino, K., Satyanarayana, S., Sitti, M.: Manufacturing of two and three-dimensional micro/nanostructures by integrating optical tweezers with chemical assembly. Robotica **23**, 435–439 (2005)
2. Antonoyiannakis, M.I., Pendry, J.B.: Electromagnetic forces in photonic crystals. Phys. Rev. B **60**, 2363–2374 (1999)

3. Labeyrie, A., Fournier, J.M., Stachnik, R.: Laser-trapped mirrors in space: Steps towards laboratory testing. Proc. SPIE **5514**, 365–370 (2004)

4. Fournier, J.M., Boer, G., Delacrétaz, G., jacquot, P., Rohner, J., Salathé, R.P.: Building optical matter with binding and trapping forces. Proc. SPIE **5514**, 309–317 (2004)

5. Liesener, J., Reicherter, M., Haist, T., Tiziani, H.J.: Multi-functional optical tweezers using computer-generated holograms. Opt. Commun. **185**, 77–82 (2000)

6. Curtis, J.E., Koss, B.A., Grier, D.G.: Dynamic holographic optical tweezers. Opt. Commun. **207**, 169–175 (2002)

7. Dholakia, K., Reece, P.: Optical micromanipulation takes hold. Nano Today **1**(1), 18–27 (2006)

8. Čižmár, T., Šiler, M., Šerý, M., Zemánek, P., Garcés-Chávez, V., Dholakia, K.: Optical sorting and detection of submicromter objects in a motional standing wave. Phys. Rev. B **74**, 035105 (2006)

9. Ashkin, A.: Acceleration and trapping of particles by radiation pressure. Phys. Rev. Lett. **24**, 156–159 (1970)

10. Ashkin, A.: Trapping of atoms by resonance radiation pressure. Phys. Rev. Lett. **40**, 729–732 (1978)

11. Burns, M.M., Fournier, J.M., Golovchenko, J.A.: Optical binding. Phys. Rev. Lett. **63**(12), 1233–1236 (1989)

12. Burns, M.M., Fournier, J.M., Golovchenko, J.A.: Optical matter: Crystalization and binding in intense optical fields. Science, **249**, 749–754 (1990)

13. Ng, J., Lin, Z.F., Chan, C.T., Sheng, P.: Photonic clusters formed by dielectric microspheres: Numerical simulations. Phys. Rev. B **72**, 085130 (2005)

14. Tatarkova, S.A., Carruthers, A.E., Dholakia, K.: One-dimensional optically bound arrays of microscopic particles. Phys. Rev. Lett. **89**(28), 283901 (2002)

15. Singer, W., Frick, M., Bernet, S., Ritsch-Marte, M.: Self-organized array of regularly spaced microbeads in a fiber-optical trap. J. Opt. Soc. Am. B **20**(7), 1568–1574 (2003)

16. Metzger, N.K., Wright, E.M., Sibbett, W., Dholakia, K.: Visualization of optical binding of microparticles using a femtosecond fiber optical trap. Optics Express **14**(8), 3677–3687 (2006)

17. Metzger, N.K., Dholakia, K., Wright, E.M.: Observation of bistability and hysteresis in optical binding of two dielectric spheres. Phys. Rev. Lett. **96**, 068102 (2006)

18. Metzger, N.K., Wright, E.M., Dholakia, K.: Theory and simulation of the bistable behaviour of optically bound particles in the Mie size regime. New J. Phys. **8**, 139 (2006)

19. Mellor, C.D., Bain, C.D.: Array formation in evanescent waves. Chem. Phys. Chem. **7**(2), 329–332 (2006)

20. Mellor, C.D., Fennerty, T.A., Bain, C.D.: Polarization effects in optically bound particle arrays. Opt. Express, **14**, 10079–10088 (2006)

21. Taylor, J.M., Wong, L.Y., Bain, C.D., Love, G.D.: Emergent properties in optically bound matter. Opt. Express **16**, 6921–6929 (2008)

22. Šiler, M., Čižmár, T., Šerý, M., Zemánek, P.: Optical forces generated by evanescent standing waves and their usage for sub-micron particle delivery. Appl. Phys. B **84**, 157–165 (2006)

23. Reece, P.J., Wright, E.M., Dholakia, K.: Experimental observation of modulation instability and optical spatial soliton arrays in soft condensed matter. Phys. Rev. Lett. **98**, 203902 (2007)

24. Kawano, M., Blakely, J.T., Gordon, R., Sinton, D.: Theory of dielectric micro-sphere dynamics in a dual-beam optical trap. Opt. Express **16**, 9306–9317 (2008)

25. Taylor, J.M., Love, G.D.: Spontaneous symmetry breaking and circulation by microparticle chains in Gaussian beam traps. Phys. Rev. A **80**, 053808 (2009)

26. Rudd, D., López-Mariscal, C., Summers, M., Shahvisi, A., Gutiérrez-Vega, J.C., McGloin, D.: Fiber based optical trapping of aerosols. Opt. Express **16**(19), 14550–14560 (2008)

27. Summers, M.D., Burnham, D.R., McGloin, D.: Trapping solid aerosols with optical tweezers: A comparison between gas and liquid phase optical traps. Opt. Express **16**, 7739–7747 (2008)

28. Guillon, M., Moine, O., Stout, B.: Longitudinal optical binding of high optical contrast microdroplets in air. Phys. Rev. Lett. **96**, 143902 (2006)
29. Guillon, M., Moine, O., Stout, B.: Erratum: Longitudinal optical binding of high contrast microdroplets in air. Phys. Rev. Lett. **99**, 079901 (2007)
30. Guillon, M., Stout, B.: Optical trapping and binding in air: Imaging and spectroscopic analysis. Phys. Rev. A **77**, 023806 (2008)
31. Ashkin, A., Dziedzic, J.M., Bjorkholm, J.E., Chu, S.: Observation of a single-beam gradient force optical trap for dielectric particles. Opt. Lett. **11**(5), 288–290 (1986)
32. McGloin, D.: Optical tweezers: 20 years on. Phil. Trans. R. Soc. A **364**, 3521–3537 (2006)
33. Visscher, K., Brakenhoff, G.J., Krol, J.J.: Micromanipulation by "multiple" optical traps created by a single fast scanning trap integrated with the bilateral confocal scanning laser microscope. Cytometry **14**, 105–114 (1993)
34. Metzger, N.K., Marchington, R.F., Mazilu, M., Smith, R.L., Dholakia, K., Wright, E.M.: Measurement of the restoring forces acting on two optically bound particles from normal mode correlations. Phys. Rev. Lett. **98**, 068102 (2007)
35. Meiners, J.C., Quake, S.R.: Direct measurement of hydrodynamic cross correlations between two particles in an external potential. Phys. Rev. Lett **82**(10), 2211–2214 (1999)
36. Garcés-Chávez, V., McGloin, D., Melville, H., Sibbett, W., Dholakia, K.: Simultaneous micromanipulation in multiple planes using a self-reconstructing light beam. Nature **419**, 145–147 (2002)
37. Karásek, V., Brzobohatý, O., Zemánek, P.: Longitudinal optical binding of several spherical particles studied by the coupled dipole method. J. Opt. A, **11**, 034009 (2009)
38. Karásek, V., Čižmár, T., Brzobohatý, O., Zemánek, P., Garcés-Chávez, V., Dholakia, K.: Long-range one-dimensional longitudinal binding. Phys. Rev. Lett. **101**, 143601 (2008)
39. Roichman, Y., Grier, D.G.: Three-dimensional holographic ring traps. Proc. SPIE **6483**, 64830F (2007)
40. McGloin, D., Carruthers, A.E., Dholakia, K., Wright, E.M.: Optically bound microscopic particles in one dimension. Phys. Rev. E **69**, 021403 (2004)
41. Mazilu, M., Dholakia, K.: Modelling the optical interactions between hundreds of micro-particles. In: Photon 08, Heriot-Watt University, UK (2008)
42. Gordon, R., Kawano, M., Blakely, J.T., Sinton, D.: Optohydrodynamic theory of particles in a dual-beam optical trap. Phys. Rev. B **77**, 245125 (2008)

Chapter 2
Scattering Theory

"It's all just a bunch of balls and sticks" [1].

2.1 Introduction

In this chapter we discuss theoretical techniques and simulation methods for calculating the scattering of a coherent laser field by one or more particles. Where possible, our treatment of Mie scattering theory uses simple linear algebra representation of the concepts involved, with the mathematical details confined to appendices. In formulating this approach we have drawn together results from a wide variety of different papers and textbooks on the subject into a coherent and clear development. The most significant novel content of this chapter is the derivations of the beam expansion coefficients for Bessel and Gaussian beams in Sects. 2.4.4 and 2.4.5, and the derivations of the expressions for the gradient potential (Sect. 2.6.2) and the pressure inside a liquid droplet (Appendix C.2).

Generalized Lorentz-Mie Theory (GLMT) is a popular technique for analyzing laser trapping and manipulation experiments. It uses an exact vector description of coherent, monochromatic light. It can be applied to particles of any size, from the Rayleigh limit up to the ray optics regime, but it is most efficient for particles in the Mie regime: particles whose size is comparable to the wavelength of the light.

The theory was first derived by Mie [2] and Debye [3] in the 1900s, and was initially restricted to plane waves. Early applications included atmospheric aerosol physics, the calculation of the angular dependency of scattering by isolated aerosol or soot particles, the optical properties of disperse colloidal solutions, and radar cross sections. The theory was subsequently generalized to treat an arbitrary beam, as laid out in 1941 by Stratton [4].

As early as 1967, Liang and Lo [5, 6] described a technique based on matrix inversion for the calculation of the field scattered by two particles. In 1988 the "order of scattering" technique (discussed later), which is critical for calculations involving more than two weakly-interacting spheres, was introduced by Fuller and Kattawar [7, 8]. A contemporary description of the mathematical tools needed to

J. M. Taylor, *Optical Binding Phenomena: Observations and Mechanisms*,
Springer Theses, DOI: 10.1007/978-3-642-21195-9_2,
© Springer-Verlag Berlin Heidelberg 2011

extend GLMT to multiple sphere configurations (which we will discuss in Sect. 2.5) can be found in [9].

Initially, in the next section, we will consider the problem in abstract terms, without concerning ourselves with the full algebraic detail of the technique. We will then revisit the technique in more detail in subsequent sections, considering scattering by a single particle (Sect. 2.3) and then the main computationally demanding steps: the expansion of the laser beam in terms of vector spherical wavefunctions (Sect. 2.4), the treatment of multiple interacting particles (Sect. 2.5), and the calculation of forces on the particles (Sect. 2.6) Following this theoretical development we will discuss the use of this calculation to determine particle motion (Sect. 2.7).

The theoretical framework of GLMT makes use of a considerable number of special functions which will be used throughout this chapter and beyond. Appendix A lists the standard symbols used for these functions, along with the symbols used to represent physical quantities. Subsequent appendices contain a number of results and derivations which have been omitted from the main text to avoid overburdening it. Finally, Appendix D discusses the implementation in computer code of the equations and techniques discussed in this chapter.

2.2 Overview

We can decompose an electromagnetic field $\mathbf{e}(\mathbf{r})$ into a complete orthonormal basis of eigenfunctions $\mathbf{e}_i(\mathbf{r})$ each with amplitude a_i:

$$\mathbf{e}(\mathbf{r}) = \sum_i a_i \mathbf{e}_i(\mathbf{r}). \tag{2.1}$$

We define the external field as the radiation field, often the field of a laser beam, in the absence of any particles. We define the incident field as the *total* incoming field impinging on the particle. In the case of a single particle the external and incident fields are the same, but the distinction is important in the case of multiple interacting particles, where scattered light from one particle will interact with the other particles and contribute to their total incident field.

A particle exposed to a given incident field will produce a scattered field, which can again be represented in that same basis, with amplitudes s_i:

$$\mathbf{s} = \mathbf{T} \cdot \mathbf{a}, \tag{2.2}$$

where $\mathbf{T}$ is a matrix describing the scattering behaviour of the particle. In general $\mathbf{T}$ will depend on the shape and physical properties of the particle, and for dielectric spheres it is determined by considering the boundary conditions on the EM field at the dielectric interface. The field that would be measured by a probe at a particular coordinate in the experiment is the sum of the external and scattered fields.

GLMT deals with spherical particles, and thus the symmetry of the problem naturally leads to the choice of the vector spherical wavefunctions (VSWFs) $\mathbf{M}_{mn}$ and $\mathbf{N}_{mn}$ for the basis $\mathbf{e}_i$ [4, 10], as they are separable into a radial and an angular part, and each individual VSWF is a solution to Maxwell's equations. Higher values of n correspond to wavefunctions whose amplitude peaks at larger radii (as we will see later when we discuss the choice of a cutoff value for n in Sect. 2.4), and higher values of $|m|$ correspond to more rapid angular variations in the function.

Although spherical wavefunctions are particularly convenient in the case of spherical particles, they form a complete basis set and hence can in principle be applied to scattering from particles of any shape. This most general approach is known as the "T-matrix" method, but here we restrict ourselves to spherical particles. In this special case the matrix $\mathbf{T}$ in (2.2) is diagonal and independent of m.

GLMT generalises easily to multiple particles due to the linearity of the electromagnetic field equations. We must simply ensure that when considering the scattering behaviour of each individual particle (determined by the surface boundary conditions) we use the *total* field incident on the particle, which is the sum of the external field and all the scattered waves from the other particles. Thus the total field $\mathbf{a}^{(k)}$ incident on particle k (as appearing in (2.2)) is:

$$\mathbf{a}^{(k)} = \mathbf{a}^{(k)}_{(ext)} + \sum_{j \neq k} \mathbf{s}'(j), \tag{2.3}$$

where $\mathbf{s}'(j)$ are the field coefficients for the field scattered by sphere j, in the basis of VSWFs *centred on particle k*.

The difficulty here is that the scattered field $\mathbf{s}^{(j)}$ is expanded in the basis of VSWFs centred on particle j, whereas here we need the field $\mathbf{s}'(j)$, expanded in the basis of VSWFs centred on particle k. We therefore need to know the matrix $\mathbf{F}$ which will transform from one basis to the other:

$$\mathbf{s}'(j) = \mathbf{F} \cdot \mathbf{s}^{(j)}. \tag{2.4}$$

Calculating the elements of this matrix is far from trivial, and will be discussed in detail in Sect. 2.5.

Equation (2.3) applies simultaneously to every particle k in the system, written in their individual bases, and the result is a large system of coupled equations which can be solved (with some computational effort) to determine the resultant field.

In the following sections we will discuss in detail the scattering by a single particle (Sect. 2.3), and then address the three computationally-challenging stages for a GLMT calculation: generating the beam shape coefficients (Sect. 2.4), treating a multi-particle system (Sect. 2.5), and calculating the force on a particle (Sect. 2.6). Following that we will discuss our actual implementation of GLMT into computer code.

2.3 Scattering by a Single Spherical Particle

Let us return for now to the case of a single particle of radius a and refractive index n_{sphere}, and a radiation field of wavenumber k in the medium of refractive index n_{ext} surrounding the particle. In this section we will expand on the abstract equations already presented, and specify the actual equations required to implement GLMT calculations.

The normalized VSWFs $\widetilde{\mathbf{M}}_{mn}$ and $\widetilde{\mathbf{N}}_{mn}$ are defined in Appendix A (the tildes emphasize that these VSWFs are normalized, in contrast to the traditional form of the VSWFs used for example in [11, 12]). The attraction of using these functions as a basis for representing the electromagnetic field in a homogeneous dielectric medium (without free charges) is that each individual function is a solution to Maxwell's equations and, since the VSWFs form a complete orthogonal set, any coherent field can be represented as a sum of normalized VSWFs. For the external electric field $\mathbf{E}$ and magnetic field $\mathbf{H}$, we use VSWFs built from spherical Bessel functions of the first kind (which is indicated by the superscript 1 as explained in Appendix A), since they are finite at the origin. The coefficients for each of these VSWFs are p_{mn} and q_{mn} :

$$
\begin{aligned}
\mathbf{E}_{ext}(k\mathbf{r}) &= -i\sum_{n=1}^{\infty}\sum_{m=-n}^{n}\left(p_{mn}\widetilde{\mathbf{N}}_{mn}^{(1)}(k\mathbf{r}) + q_{mn}\widetilde{\mathbf{M}}_{mn}^{(1)}(k\mathbf{r})\right), \\
\mathbf{H}_{ext}(k\mathbf{r}) &= -n_{ext}\sum_{n=1}^{\infty}\sum_{m=-n}^{n}\left(q_{mn}\widetilde{\mathbf{N}}_{mn}^{(1)}(k\mathbf{r}) + p_{mn}\widetilde{\mathbf{M}}_{mn}^{(1)}(k\mathbf{r})\right).
\end{aligned}
\tag{2.5}
$$

The expression for the scattered field (the field scattered by the particle) expanded in terms of the coefficients a_{mn} and b_{mn} is similar, except we use VSWFs built from spherical Hankel functions of the first kind, since in the far-field limit they describe outgoing spherical waves:

$$
\begin{aligned}
\mathbf{E}_{scat}(k\mathbf{r}) &= i\sum_{n=1}^{\infty}\sum_{m=-n}^{n}\left(a_{mn}\widetilde{\mathbf{N}}_{mn}^{(3)}(k\mathbf{r}) + b_{mn}\widetilde{\mathbf{M}}_{mn}^{(3)}(k\mathbf{r})\right), \\
\mathbf{H}_{scat}(k\mathbf{r}) &= n_{ext}\sum_{n=1}^{\infty}\sum_{m=-n}^{n}\left(b_{mn}\widetilde{\mathbf{N}}_{mn}^{(3)}(k\mathbf{r}) + a_{mn}\widetilde{\mathbf{M}}_{mn}^{(3)}(k\mathbf{r})\right).
\end{aligned}
\tag{2.6}
$$

The expression for the internal field (the field internal to the particle) is very similar to that for the external field, but note the change of sign and the fact that the VSWFs are evaluated as a function of the position coordinate $k'\mathbf{r}$, where k' is the wavenumber *inside* the particle, $k' = \frac{n_{int}}{n_{ext}}k$, which can be complex if the internal refractive index n_{int} is complex:

$$
\begin{aligned}
\mathbf{E}_{int}(k\mathbf{r}) &= -i\sum_{n=1}^{\infty}\sum_{m=-n}^{n}\left(d_{mn}\widetilde{\mathbf{N}}_{mn}^{(1)}(k'\mathbf{r}) + c_{mn}\widetilde{\mathbf{M}}_{mn}^{(1)}(k'\mathbf{r})\right), \\
\mathbf{H}_{int}(k\mathbf{r}) &= -n_{ext}\sum_{n=1}^{\infty}\sum_{m=-n}^{n}\left(c_{mn}\widetilde{\mathbf{N}}_{mn}^{(1)}(k'\mathbf{r}) + d_{mn}\widetilde{\mathbf{M}}_{mn}^{(1)}(k'\mathbf{r})\right).
\end{aligned}
\tag{2.7}
$$

It is worth noting at this point that a number of different conventions exist for the exact form of the beam expansion. For example Barton [13] and Čizmar[12] use a slightly different version of (2.5). For example, their external field coefficients A_{lm} and B_{lm} are related to our p_{mn} and q_{mn} as follows:

$$A_{lm} = \frac{ip_{mn}}{2\pi(ka)},$$
$$B_{lm} = \frac{n_{ext}q_{mn}}{2\pi(ka)}.$$

(2.8)

One particularly important convention is whether the coefficients p_{mn}, q_{mn} etc. are normalized. Several authors such as Mackowski and Xu [14] do not normalize them. While this remains closer to the original Mie theory, it poses serious problems for spheres tens of wavelengths in diameter, and for clusters of spheres. The coefficients grow combinatorially with m and n, and they can overflow the variables used to store them in a computer program. With the scheme presented here, the beam shape coefficients are all of order 1, and the magnitudes of $p_{\pm mn}$ and $q_{\pm mn}$ are all the same (for an axial beam). They are also independent of the size of the sphere, which seems more appropriate since they describe the intrinsic properties of the laser beam, and so conceptually should not depend on the radius of the sphere they are centred on.

Now that we have defined a basis for the incident and scattered fields, we must determine how they are related. The scattered field is uniquely determined by the field incident on the surface of the particle, and it must be such that the electric and magnetic boundary conditions at a dielectric interface are satisfied by the total external field and the field internal to the particle [15]. The scattering matrix **T** referred to in (2.2) must be derived in such a way as to satisfy those surface boundary conditions.

In the case of a sphere, **T** is diagonal and independent of m, and $a_{mn} = \alpha_n p_{mn}$ and $b_{mn} = \beta_n q_{mn}$. The derivation for the scattering coefficients α_n and β_n is well known [16], but numerical overflow can be an issue when working with very large spheres if the resultant formula is used naively. A particularly convenient form for the equations exists which solves this problem, involving logarithmic derivatives of the Riccati-Bessel functions [17], along with appropriate recurrence formulae to reliably generate the coefficients:

$$\alpha_n = \frac{\psi_n(x)}{\zeta_n(x)}\left(\frac{D_n(mx) - mD_n(x)}{D_n(mx) - mG_n(x)}\right),$$
$$\beta_n = \frac{\psi_n(x)}{\zeta_n(x)}\left(\frac{mD_n(mx) - D_n(x)}{mD_n(mx) - G_n(x)}\right),$$

(2.9)

where

$$D_n(z) = [\ln \psi_n(z)]',$$
$$G_n(z) = [\ln \zeta_n(z)]',$$
$$\psi_n(z) = zj_n(z),$$
$$\zeta_n(z) = zh_n^{(1)}(z),$$
$$m = n_{sphere}/n_{ext}.$$

We have now defined a basis for representing the electromagnetic field inside and outside a dielectric particle, and know how to relate the scattered field to the incident field. Once we know how to expand the external field in terms of the beam shape coefficients p_{mn} and q_{mn} (which will be discussed in the next section), then we can calculate the field scattered by any single particle exposed to a laser field. Quantities such as the scattering cross section or far-field scattering pattern can then be determined [16]. One quantity of particular interest for optical trapping and manipulation is the force exerted on the particle by the radiation field. How this is calculated will be discussed in Sect. 2.6.

2.4 Beam Representations

For a practical calculation it is necessary to truncate the sum over VSWFs (2.5) at some finite $n = n_{max}$. The choice of this cutoff value is generally taken as:

$$n_{max} = (ka) + 4.05(ka)^{1/3} + 2. \tag{2.10}$$

This is a purely empirical result that was originally proposed by Wiscombe [18]. In the case of very strong resonances[1] it can be necessary to increase this value slightly, but for off-resonance calculations it is sufficient. If there is concern as to whether a calculation is accurate, it can be repeated with a slightly higher n_{max} to see whether the same result is obtained.

This cutoff can be understood in a number of ways. An arbitrary field contained within a radius a can be extremely closely approximated using only those VSWFs with $n \leq n_{max}$. Spherical bessel functions of order $n > n_{max}$ are very close to zero for $r \leq a$, which means that they have negligible effect on the field on the surface of the sphere. The field at a larger distances from the sphere will not in general be well described for a given choice of the cutoff n_{max}, but the field at that point distant from the sphere has no effect on the scattering behaviour of the sphere, and so *it is not a problem* that we have not correctly described the field at that point.

Alternatively (for the same mathematical reasons as in the above argument), the scattering coefficients of a sphere of radius a are negligible above that value of n_{max}. The fact that beyond that point they decay very rapidly with n is a practical reason why there is no need to calculate the beam shape coefficients beyond that value of n: mathematically they will not have any effect on the scattered coefficients.

This cutoff effect is illustrated in Fig. 2.1. It can be seen that as n_{max} is increased, the field represented by the VSWF coefficients up to the cutoff $n = n_{max}$ provides a good representation of the plane wave over a broader region centred on the particle. The particle has a size parameter of 10, for which (2.10) demands a

---

[1] Resonances include "whispering gallery modes", where standing waves develop around the internal surface of a single sphere, and cavity-like resonances between two nearby spheres.

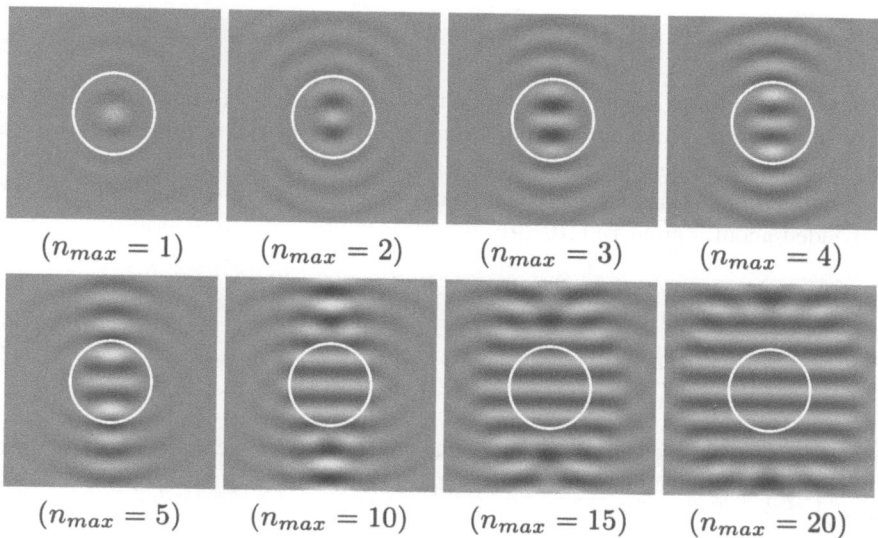

$(n_{max} = 1)$ $(n_{max} = 2)$ $(n_{max} = 3)$ $(n_{max} = 4)$

$(n_{max} = 5)$ $(n_{max} = 10)$ $(n_{max} = 15)$ $(n_{max} = 20)$

Fig. 2.1 Representation of a plane wave with increasing numbers of VSWFs, up to $n = n_{max}$. As n_{max} increases, the plane wave is correctly represented over a wider and wider area centred on the particle (whose surface is indicated by the *white circle*)

value of $n_{max} = 21$. Notice that even by $n = 10$ the field is visually very close to the required plane wave distribution over the surface of the sphere.

2.4.1 Integral Method

The most general way of calculating the beam coefficients is to decompose the beam function into VSWFs using an orthogonal eigenfunction transform (a generalization of the Fourier transform which exploits the orthogonal nature of the VSWFs to obtain the p_{mn} and q_{mn} expansion coefficients):

$$p_{mn} = i \times \frac{\int_S \mathbf{E} \cdot \widetilde{\mathbf{N}}_{mn}^* \mathrm{d}S}{\int_S \left| \widetilde{\mathbf{N}}_{mn} \right|^2 \mathrm{d}S}, \qquad (2.11)$$

and similarly for q_{mn}, using $\widetilde{\mathbf{M}}_{mn}$ instead of $\widetilde{\mathbf{N}}_{mn}$.

This calculation can be performed for an arbitrary beam, but that requires the evaluation of $O(n_{max}^2)$ surface integrals, which makes it very slow compared to the rest of the Mie calculation. Fortunately this integral can be fully or partially solved for commonly-used beam types, as will be outlined below.

2.4.2 Plane Wave

While a true, infinite plane wave is not physically realistic, it is a model commonly used in scattering calculations. For example, a very broad, collimated Gaussian beam can be locally approximated by a plane wave.

Equation (2.11) has an analytical solution in the case of a plane wave $\mathbf{e}_0 e^{i\mathbf{k}\cdot\mathbf{r}}$ expanded about a point $\mathbf{r}_0$ [10, 19]:

$$
\left\{ \begin{array}{c} p_{mn} \\ q_{mn} \end{array} \right\} = U_{mn} \left(e_\theta \left\{ \begin{array}{c} \tilde{\tau}_{mn}(\cos\theta) \\ \tilde{\pi}_{mn}(\cos\theta) \end{array} \right\} - i e_\theta \left\{ \begin{array}{c} \tilde{\pi}_{mn}(\cos\theta) \\ \tilde{\tau}_{mn}(\cos\theta) \end{array} \right\} \right) e^{-im\phi} e^{i\mathbf{k}\cdot\mathbf{r}_0},
$$
$$
U_{mn} = \frac{4\pi i^n}{n(n+1)},
$$

(2.12)

where θ and ϕ are the zenith and azimuthal angles of the wavenumber $\mathbf{k}$. Hence we can analytically calculate the beam shape coefficients p_{mn} and q_{mn} for a plane wave.

2.4.3 Evanescent Wave

When a plane wave encounters a planar dielectric boundary from a medium of refractive index n_{subst} to a medium of refractive index n_{ext}, it is refracted according to Snell's law which relates the angle of refraction θ_r to angle of incidence θ_i:

$$
n_{ext} \sin\theta_r = n_{subst} \sin\theta_i.
$$

(2.13)

If the wave in the xz plane is incident on an interface at x = 0 then the wavevector in the second medium will be $\mathbf{k} = k(\cos\theta_r, 0, \sin\theta_r)$ in cartesian coordinates.

If the refractive index mismatch is sufficiently large, we may find that $\sin\theta_r$ is greater than 1. This is unphysical for real θ_r, and the wave is totally internally reflected at the boundary. However we can still satisfy Snell's law if we allow θ_r to be imaginary. In this case $\cos\theta_r$ will be imaginary, and so the x component of the wavevector will be imaginary. This represents the exponential decay of a wave in the x direction. This is known as the evanescent wave. In the case of a perfect plane interface between two infinite homogeneous media, no energy flows through the interface,[2] although the field on the other side of the interface is not zero. The field decays over a distance comparable to the wavelength of the light, and the decay length is greatest at the critical angle where total internal reflection first occurs.

Figure 2.2 shows the behaviour of a wave above and below the critical angle. The evanescent wave can be considered entirely equivalent to the standard plane

---

[2] As we will see in Chap. 3, however, the presence of a particle on the far side of the interface will change the boundary conditions and cause energy and momentum to be transferred across the boundary.

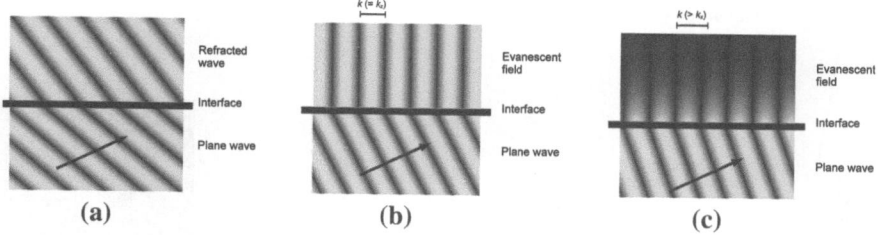

Fig. 2.2 Diagrams showing the refractive behaviour of a plane wave at a dielectric interface (*reflected wave not shown*): **a** wave is refracted into the upper medium; **b** at the critical angle, the wave is refracted parallel to the boundary; **c** beyond the critical angle (*complex angle of refraction*) the wave is totally internally reflected and an evanescent field develops on the other side of the boundary

wave transmitted below the critical angle, but with an imaginary wavevector as stated above [20]. This wavevector can be used in (2.12) to calculate the beam shape coefficients for an evanescent wave.

2.4.4 Bessel Beam

A Bessel beam is a non-diffracting beam which is of interest in optical confinement and guiding experiments [21, 22]. An ideal Bessel beam is formed from a conical spectrum of converging plane waves, and in practical terms a Bessel beam can be generated by passing a Gaussian beam through an axicon (a conical lens). Figure 2.3 shows a schematic diagram of an axicon bring used to realize a Bessel beam. Čizmar et al. [12] considered the VSWF expansion coefficients of a Bessel beam, but their expression required the numerical evaluation of an integral, which can end up being the bottleneck in the overall Mie calculation in the case where a single sphere is treated. In this section we derive an analytical result for the VSWF expansion coefficients of a Bessel beam.

Consider a particle at position $\mathbf{r}_0 = (x, y, z)$ which is exposed to the field of an x-polarized Bessel beam propagating along the z axis. We represent the beam by a sum of plane waves $\mathbf{e}_0 e^{i\mathbf{k}\cdot\mathbf{r}}$ making an angle α_0 with the z axis [12], as illustrated in Fig. 2.3

$$E(\mathbf{r}) = E_0 \int_0^{2\pi} \mathbf{e}_0(\alpha_0, \beta) e^{i\mathbf{k}\cdot\mathbf{r}} \mathrm{d}\beta, \qquad (2.14)$$

where $\mathbf{k} = (k, \alpha_0, \beta)$ (in polar coordinates), and $\mathbf{e}_0 = \cos\beta \mathbf{i}_\theta - \sin\beta \mathbf{i}_\phi$. Our equation is equivalent to Čizmar's Equation. (A.1), but note that we have defined β slightly differently in order to arrive at a nearer final result. Hence the plane wave polarization $\mathbf{e} = \mathbf{e}(\mathbf{k})$ is

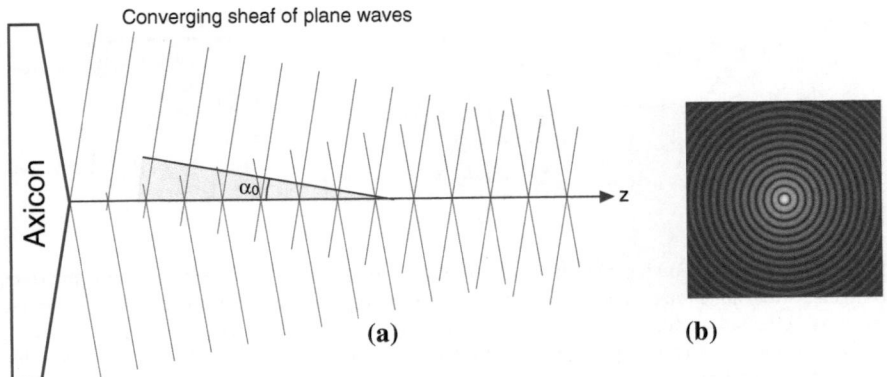

Fig. 2.3 **a** Schematic diagram of an axicon being used to realize a Bessel beam. α_0 is the angle between the direction of propagation of the plane waves and the z axis (*direction of beam propagation*). **b** Cross-section through a Bessel beam (*logarithm of intensity*)

$$
\begin{aligned}
\mathbf{e}_x &= \cos\alpha_0 + \sin^2\beta(1 - \cos\alpha_0) \\
\mathbf{e}_y &= -[(1 - \cos\alpha_0)\sin\beta\cos\beta] \\
\mathbf{e}_z &= -\sin\alpha_0\cos\beta,
\end{aligned}
\tag{2.15}
$$

with the opposite sign to Cizmar's Equation (A.3) on $\mathbf{e}_z$, in consequence of our different β. Note that it can of course be verified that $\mathbf{k}\cdot\mathbf{e} = 0$. $\mathbf{e}$ can be represented in the spherical basis of $\mathbf{k}$ using the transformation:

$$
\begin{aligned}
\begin{pmatrix} \mathbf{e}_r \\ \mathbf{e}_\theta \\ \mathbf{e}_\phi \end{pmatrix}
&= \begin{pmatrix}
\cos\beta\sin\alpha_0 & \sin\beta\sin\alpha_0 & \cos\alpha_0 \\
\cos\beta\cos\alpha_0 & \sin\beta\cos\alpha_0 & -\sin\alpha_0 \\
-\sin\beta & \cos\beta & 0
\end{pmatrix} \mathbf{E}_{PW}^{(cart)} \\
&= \begin{pmatrix} 0 \\ \cos\beta \\ -\sin\beta \end{pmatrix}
\end{aligned}
\tag{2.16}
$$

$\mathbf{e}_\theta = \cos(\beta)$ is the component of the polarization vector $\mathbf{e}$ which lies in the plane containing $\mathbf{k}$ and the z axis. $\mathbf{e}_\phi = -\sin(\beta)$ is the component of $\mathbf{e}$ which is perpendicular to that plane. $\mathbf{e}_r$, the component of $\mathbf{e}$ parallel to $\mathbf{k}$, is of course zero as $\mathbf{e}\cdot\mathbf{k} = 0$ for a plane wave.

The individual plane wave expansion is given in (2.12). We substitute that plane wave expansion into (2.14), giving

$$
\begin{Bmatrix} p_{mn} \\ q_{mn} \end{Bmatrix} = E_0 U_n \times \int_0^{2\pi} \left[\cos(\phi) \begin{Bmatrix} \widetilde{\tau}_{mn} \\ \widetilde{\pi}_{mn} \end{Bmatrix} + i\sin(\phi) \begin{Bmatrix} \widetilde{\pi}_{mn} \\ \widetilde{\tau}_{mn} \end{Bmatrix} \right] e^{-im\phi} e^{i\mathbf{k}\cdot\mathbf{r}_0}\, d\phi.
\tag{2.17}
$$

The $e^{i\mathbf{k}\cdot\mathbf{r}_0}$ term can be expanded, and sine and cosine rewritten in terms of exponentials, for $\rho = k\sqrt{x^2 + y^2}\sin\theta$ and $\phi_0 = \arctan(-y/x) - \frac{\pi}{2}$ to obtain

$$\begin{Bmatrix} P_{mn} \\ q_{mn} \end{Bmatrix} = E_0 U_n e^{ikz\cos\theta} \times \left[\begin{Bmatrix} \tilde{\tau}_{mn} \\ \tilde{\pi}_{mn} \end{Bmatrix} I^{(+)} + \begin{Bmatrix} \tilde{\pi}_{mn} \\ \tilde{\tau}_{mn} \end{Bmatrix} I^{(-)} \right],$$

$$I^{(\pm)} = \frac{1}{2}\int_0^{2\pi} e^{i(1-m)\phi} e^{i\rho\cos\left(\phi+\phi_0+\frac{\pi}{2}\right)}\mathrm{d}\phi \qquad (2.18)$$

$$\pm \frac{1}{2}\int_0^{2\pi} e^{i(-1-m)\phi} e^{i\rho\cos\left(\phi+\phi_0+\frac{\pi}{2}\right)}\mathrm{d}\phi.$$

The azimuthal integral can be solved [23, (9.1.21)] in terms of Bessel functions of the first kind to give

$$I^{(\pm)} = \pi\left(e^{i(m-1)\phi_0}J_{1-m}(\rho) \pm e^{i(m+1)\phi_0}J_{-1-m}(\rho)\right). \qquad (2.19)$$

The field described by (2.5) and (2.18) has been verified to be the same as the explicit integral in (2.14). With the integral eliminated, the beam shape coefficients (BSCs) p_{mn} and q_{mn} can be calculated analytically in a fraction of the time required by existing published methods.

2.4.5 Gaussian Beam

Approaches for representing a loosely-focused Gaussian beam (empirically, for a numerical aperture $NA \lesssim 0.25$) have been known for some time. It is possible to decompose the Gaussian beam into a superposition of plane waves [24] in the near-paraxial regime. The number of plane waves required to accurately represent the beam increases as the NA increases, and the approach eventually breaks down as the NA rises, when vectorial effects not accounted for in such paraxial-type approximations become important.

Representing a tightly-focused Gaussian beam presents challenges, and much effort has been devoted to increasingly detailed nth order approximation to the beam [25–27], which could then for example be inserted into the integral of (2.11) to obtain a VSWF expansion for the beam. Such a perturbative approach cannot be applied to optical tweezers, though, since the numerical aperture (N.A.) can be greater than 1. It is however possible to precisely describe the *far-field* distribution even of a tightly-focused Gaussian beam, and Nieminen et al. [28] exploit this to build a system of linear equations which can be solved to obtain the BSCs. They define a series of discrete points in the far field, and at each point they build an equation which relates the far-field intensity to the BSCs. The relevant equation for a point $\mathbf{r}$ at which the far-field Gaussian electric field is $\mathbf{E}$ is simply (2.5), which provides two equations (the two non-zero (transverse) components of $\mathbf{E}$) in $4n_{max}$ unknowns (for the on-axis case with the beam propagating along the z axis, so that only the $m = \pm 1$ harmonics are nonzero): in the equation, every one of the p_{mn} and q_{mn} coefficients is unknown. We therefore require $2n_{max}$ such points in order to fully determine the equation system.

This linear equation approach is fairly efficient for a particle at the focus of an extremely tightly-focused beam, where the required value of n_{max} is small. However, it requires the use of translation matrices to represent particles at positions other than the focus. In addition to this, it is entirely unsuited to more loosely-focused beams, and there is an awkward transition region as a function of the N.A. where neither this approach nor the plane wave superposition approach produce satisfactory results without prohibitively long calculation times.

We took a slightly different approach, with the aim of deriving a single expression which is valid all the way from the plane wave limit to the high NA limit. The starting point was to select a suitable expression for a fully-vectorial Gaussian beam in the far field [29]:

$$\mathbf{E} = \frac{E_0 k^2 w_0^2}{2ikR} \sqrt{\cos\theta} \, e^{-(\gamma \sin\theta)^2} e^{ikR} e^{i\mathbf{r}.\mathbf{i}_r} \left(\cos\phi \mathbf{i}_\theta - \sin\phi \mathbf{i}_\phi \right), \qquad (2.20)$$

where γ is the ratio of the focal length of the objective to the beam waist size of the (broad) Gaussian beam *before* the objective. Here we have selected the global amplitude and phase to be consistent with the plane wave result in (2.12). It is important to emphasize that this is an *exact* vectorial expression; in the far field of the particle (i.e. far from the beam focus), a Gaussian beam is a transverse wave, however tightly focused it is, and however important vectorial effects are close to the focus. If the appropriate VSWF expansion coefficients can be determined by some means or other, the correct vector field close to the focus will be reproduced (even though no exact analytical form is known for that field as a function of $\mathbf{r}$), since fixing an electromagnetic field over a closed surface is enough to determine its value throughout all space.

This expression was then substituted into the general surface integral of (2.11). It is then possible to (at least partially) solve the integral by exploiting the orthogonality of the vector spherical harmonics (see Appendix A). For mathematical convenience, we elect to evaluate the p_{mn} surface integral at radius $kR = (2N\pi + \frac{n\pi}{2})$ and the q_{mn} surface integral at radius $kR = (2N\pi + \frac{(n+1)\pi}{2})$, in the limit of large N. Under these conditions, the far-field vector spherical harmonics are:

$$\sqrt{\frac{2n+1}{4\pi}\frac{(n-m)!}{(n-m)!}} \left\{ \begin{matrix} \tilde{\mathbf{N}}_{mn}^{(1)} \\ \tilde{\mathbf{M}}_{mn}^{(1)} \end{matrix} \right\} = \left(\left\{ \begin{matrix} \tilde{\tau}_{mn} \\ i\tilde{\pi}_{mn} \end{matrix} \right\} \mathbf{i}_\theta + \left\{ \begin{matrix} i\tilde{\pi}_{mn} \\ -\tilde{\tau}_{mn} \end{matrix} \right\} i_\phi \right) \frac{e^{im\phi}}{kR}. \qquad (2.21)$$

Substituting these into (2.11) and, as we did for the Bessel beam, expressing the azimuthal integral in terms of $I^\pm$, which we solved in (2.19), we find:

$$\begin{Bmatrix} p_{mn} \\ q_{mn} \end{Bmatrix} = U_n \int_0^{\theta_0} E'(\theta) \left(\begin{Bmatrix} \tilde{\tau}_{mn} \\ \tilde{\pi}_{mn} \end{Bmatrix} I^{(+)} + \begin{Bmatrix} \tilde{\pi}_{mn} \\ \tilde{\tau}_{mn} \end{Bmatrix} I^{(-)} \right) \sin\theta d\theta, \qquad (2.22)$$

where $E'(\theta) = \frac{E_0 k^2 w_0^2}{4\pi} \sqrt{\cos\theta} e^{-(\gamma\sin\theta)^2} e^{ikz\cos\theta}$ and θ_0 is the half-angle subtended by the objective at the beam focus. The θ integral cannot unfortunately be solved analytically, but we have succeeded in deriving a formula containing just a single integral, which is valid for arbitrary NA. One useful side-effect of the θ integral remaining in the final expression is that the envelope $E'(\theta)$ can be modified in any way that is desired, and hence it is trivial to impose radially-symmetric modifications to the beam profile, such as apodization.

It subsequently became clear that this was not an entirely novel result: a number of other authors have quoted similar results. The result is analogous to that derived by Doicu & Wriedt for the limit of low N.A. [25, Eq. 21], but the result here is valid for tightly-focused beams as well. Our result is also consistent with that quoted but not derived by Mazolli et al. for a circularly-polarized beam [30, 31]. We compare the Gaussian and Bessel beam expansions derived here with the performance of rival approaches in [32].

2.5 Multiple Particles

As mentioned in the overview of GLMT in Sect. 2.2, in the case of multiple particles we must solve a system of coupled equations which arise from (2.3) and (2.4). For illustration, consider a cluster of three nearby particles. We can express (2.3) for the whole system in terms of a block matrix:

$$\begin{bmatrix} \mathbf{a}^{(1)} \\ \mathbf{a}^{(2)} \\ \mathbf{a}^{(3)} \end{bmatrix} = \begin{bmatrix} \mathbf{a}_{ext}^{(1)} \\ \mathbf{a}_{ext}^{(2)} \\ \mathbf{a}_{ext}^{(3)} \end{bmatrix} + \begin{bmatrix} 0 & \mathbf{F}_{21} & \mathbf{F}_{31} \\ \mathbf{F}_{12} & 0 & \mathbf{F}_{32} \\ \mathbf{F}_{13} & \mathbf{F}_{23} & 0 \end{bmatrix} \begin{bmatrix} \mathbf{s}^{(1)} \\ \mathbf{s}^{(2)} \\ \mathbf{s}^{(3)} \end{bmatrix} \qquad (2.23)$$

$$\text{or} \quad \underline{a} \quad = \quad \underline{a}_{ext} + \qquad \underline{\underline{F}} \qquad \underline{s}$$

where $\mathbf{F}_{ij}$ is the translation matrix of (2.4) which transforms a scattered field in the basis of sphere i into an incident field in the basis of sphere j.

Before we consider the details of multiple scattering calculations, we should revisit the discussion of the determination of n_{max} (2.10) in the context of multiple particles. We need to consider whether this cutoff is still appropriate. At first glance it seems that it might not be, since the size of the cluster is larger than the size of the individual spheres, which was what we based our cutoff on.

The scattered field from a particle A can be fully described using $n \leq n_{max}$. There is no approximation here: the field sources (the surface of the sphere) are

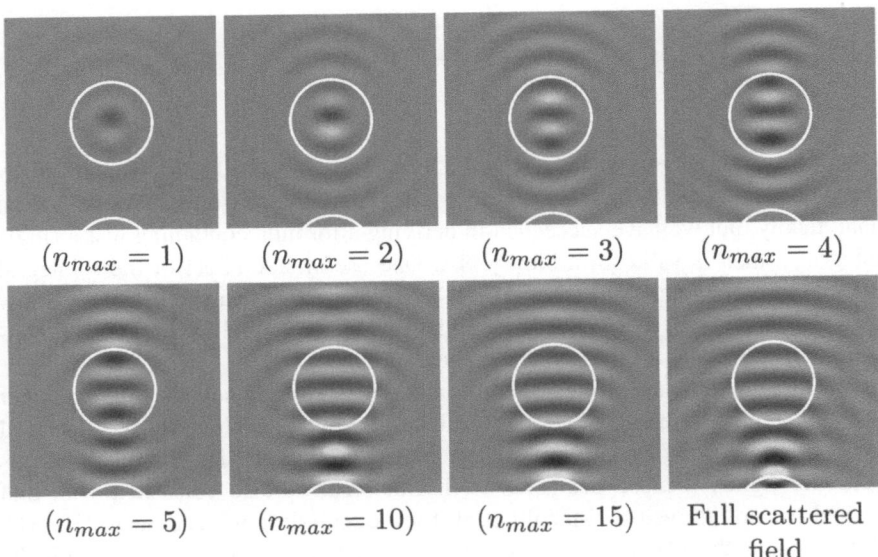

$(n_{max} = 1)$ $\qquad$ $(n_{max} = 2)$ $\qquad$ $(n_{max} = 3)$ $\qquad$ $(n_{max} = 4)$

$(n_{max} = 5)$ $\qquad$ $(n_{max} = 10)$ $\qquad$ $(n_{max} = 15)$ $\qquad$ Full scattered
field

Fig. 2.4 Scattered field from particle A (*at bottom of field of view*) represented in a basis centred around particle B (*center of field of view*). As increasing numbers of VSWFs are used (increasing n_{max}), the scattered field is correctly represented over a wider area centred on particle B. By $n_{max} = 10$ the scattered field is well represented on the surface of particle B. This cutoff is sufficient for an accurate calculation of the second-order field scattered from surface of B. When processing sphere B in this way, *it does not matter* that the field elsewhere, including on the surface of particle A, is not correctly represented in this basis

contained within that radius a of the sphere, and so the field is fully described over *all space* using this cutoff.

If we wanted to fully describe this scattered field at all points in space, but in a basis centred on another particle B at distance d away from A, then we would need to use a larger cutoff which is a function of d. However fortunately we do not need a full description of this field. When calculating the field re-scattered by particle B, we only actually need to know the field *on the surface* of that particle B. Hence the same cutoff n_{max} as we would use in the single-sphere case can be used in all parts of the calculation. Figure 2.4 shows the scattered field from particle A represented in a basis centred on particle B. In the same way as with Fig. 2.1, the field is shown for various values of n_{max}. It can be seen that by $n_{max} = 10$ the scattered field is well represented *within the volume of sphere B*, even though the scattered field at other locations, such as on the surface of sphere A, is *not* accurately described at this value of n_{max}.

In the following sections we will discuss techniques available for solving (2.23), and will then present the method for generating the translation matrices.

2.5.1 Multiple Scattering Principle

For a group of particles that are not extremely close to one another, the effects of multiple scattering are normally small. That is, the solution for the EM field is fairly well approximated by the sum of the external field and the field that would be scattered by each particle individually in the absence of the other particles. The effects of multiple scattering, where the scattered light from one particle is re-scattered by another particle, are only a small perturbation to the first-order solution.

In those circumstances, an interactive solution can normally be obtained by treating the scattered field as a perturbation to the background light field [7, 8]. We can draw a parallel between this method and the way the system would evolve if we were able to instantly "turn on" the laser field across all space. When the particles first experience the field, they scatter some light. After a very short interval that scattered light will hit the other particles in the cluster, and will be re-scattered by them. That re-scattered light will propagate out and hit the other particles, and so on.

The zero-order solution for the scattered field $\underline{s}_0$ is the field obtained by treating every sphere as a scatterer in isolation. Using the notation of (2.23):

$$\underline{s}_{(0)} = \underline{T} \cdot \underline{a}_{ext}. \qquad (2.24)$$

We then calculate the net incident field due to the zero-order scattered field:

$$\underline{a}_{(1)} = \underline{a}_{ext} + \underline{F} \cdot \underline{s}_{(0)}, \qquad (2.25)$$

where $\underline{F}$ is the translation matrix (see (2.4).The first-order scattered field is then:

$$\underline{s}_{(1)} = \underline{T} \cdot \underline{a}_{(1)}, \qquad (2.26)$$

the second-order net incident field is:

$$\underline{a}_{(2)} = \underline{a}_{ext} + \underline{F} \cdot \underline{s}_{(1)}. \qquad (2.27)$$

and so on until the solution converges.

This method can alternatively be written in the form of an infinite sum:

$$\underline{s} = [\underline{T} + (\underline{T} \cdot \underline{F} \cdot \underline{T}) + (\underline{T} \cdot \underline{F} \cdot \underline{T} \cdot \underline{F} \cdot \underline{T})] \cdot \underline{a}_{ext} \qquad (2.28)$$

Once the iterative process has been continued until the scattered field does not change significantly, a field has been obtained that satisfies (2.23). Empirically, a solution accurate to around one part in 10^{-5} can be obtained within about 10–20 iterations, even for particles whose centres are only 3 radii apart. In practice, what we are normally interested in is the force on the particles, so an appropriate convergence condition is when the force has stabilized to a suitable tolerance. Given assumptions inherent in the model, such as maybe neglecting aberrations in the optics used to generate the beams, there is not normally going to be an

accuracy gain in having an excessively strict convergence limit, and in most cases
we conservatively consider the solution to be converged if the force changes by
less than 10^{-5} between successive iterations.

Xu recommends a modified scheme where the actual nth order solution is
calculated by combining the raw nth order solution and the $(n - 1)$th order solution
using a damping factor u [14, Eq. 35]:

$$\underline{a}_{(n)} = \underline{a}_{ext} + u\underline{F} \cdot \underline{s}_{(n-1)} + (1 - u)\underline{a}_{(n-1)}. \tag{2.29}$$

Empirically we found that although varying u affected the speed of convergence, it
rarely allowed easy convergence in cases where divergence occurs for $u = 1$, and
so it is simpler just to set $u = 1$. Fortunately in most cases we were interested in,
divergence was not a problem. In cases where it is, an alternative technique can be
used which is described in the Sect. 2.5.2.

2.5.1.1 Forward- and Back-Scattering

As we will see in Chap. 4, it can sometimes be useful to separate the effects of
forward- and back-scattering in order to understand the effects at work in a system
of particles. Fortunately there is a very simple modification which can be made to
the scattering calculation (a modification which we described in [33]) which
allows us to "disable" the effects of back-scattering.

Recall the concise representation of the scattering interaction in (2.23), which
for a given particle i can be written:

$$\mathbf{a}^{(i)} = \mathbf{a}_{ext}^{(i)} + \sum_{j \neq i} \mathbf{F}_{ji}.\mathbf{s}^{(j)}.$$

It is easy to alter this calculation so that backscatter is not taken into account. If the
particles are indexed "upstream" to "downstream" (i.e. with the first particle
closest to the laser source), then we simply modify the sum to read:

$$\mathbf{a}^{(i)} = \mathbf{a}_{ext}^{(i)} + \sum_{j < i} \mathbf{F}_{ji}.\mathbf{s}^{(j)}$$

Although such a model is un-physical, there can be significant benefits in
modeling such a situation: a real-world experiment will be subject to Brownian
motion of the trapped particles, which will wash out small local minima in the
interactions caused by the back-scattered field. In the absence of Brownian
motion (i.e. at absolute zero), a simulation can easily become trapped in such a
local minimum. Since it is computationally very expensive to include
Brownian motion in a simulation, choosing a model which does not include
the back-scattered light is a useful compromise in many cases (though the

results of these un-physical simulations must be verified using a full physically-realistic model).

2.5.2 Solution by Inversion

Where the multiple-scattering technique described in the previous section breaks down, a solution to Maxwell's equations which satisfies all the boundary conditions does still of course exist. It could in principle be arrived at by an alternative physically-inspired method: possibly by a gradual increase of the refractive index of the particles, starting from that of the surrounding medium. However it is more practical to directly solve the system of (2.3) by inversion [5, 6]. For N particles and a cutoff value for the summation of VSWFs at n_{max}, we are presented with simultaneous equations in $2N(n_{max}^2 + 2n_{max})$ unknowns. This will be computationally very difficult to solve for any reasonably large value of n_{max} as the time required for the solution will scale as $O(N^3 n_{max}^6)$ [8].

Fortunately in the case of spheres lying on the z axis a simplification exists. As will be discussed in Sect. 2.5.3, a translation matrix $\mathbf{F}$ along the z axis does not connect coefficients of different m (the relevant matrix elements are zero). Thus our system of equations can be decomposed into $(2n_{max} + 1)$ independent systems of order $(N \times n_{max})$ unknowns, which can be solved in $O(N^3 \times n_{max}^4)$ time. Furthermore, the fact that there are $(2n_{max} + 1)$ completely independent systems means the problem is naturally suited to calculation on parallel computing architectures for increased speed.

When we are considering two spheres, as is the case in [34] for example, the axes can always be chosen so that the spheres both lie on the z axis, and thus the system can always be solved using the z axis simplification.

Due to the ill-conditioned characteristics of the matrices involved, and the large contrast in magnitude of the various terms in the matrix, standard techniques for solving the equations such as LU decomposition tend to amplify small errors, caused for example by rounding-off errors, and do not generate a good solution to the equations. The "generalised minimal residual method" [35], however, is found to give reliable solutions.

2.5.3 Translation Matrix Simplification

We require a translation matrix (as used schematically in (2.25) in order to switch between a VSWF basis centred on one particle to a basis centred on a different particle. The following equation states the addition theorem for VSWFs, expanding a single VSWF centred at the point $\mathbf{r}$ in terms of an infinite sum of VSWFs centred at a different point $\mathbf{r}'$ [9]:

$$\widetilde{\mathbf{M}}_{mn}^{(3)}(k\mathbf{r}) = \sum_{l=1}^{\infty} \sum_{k=-l}^{l} \left[A_{kl}^{mn} \widetilde{\mathbf{M}}_{kl}^{(1)}(k\mathbf{r}') + B_{kl}^{mn} \widetilde{\mathbf{N}}_{kl}^{(1)}(k\mathbf{r}') \right],$$

$$\widetilde{\mathbf{N}}_{mn}^{(3)}(k\mathbf{r}) = \sum_{l=1}^{\infty} \sum_{k=-l}^{l} \left[A_{kl}^{mn} \widetilde{\mathbf{N}}_{kl}^{(1)}(k\mathbf{r}') + B_{kl}^{mn} \widetilde{\mathbf{M}}_{kl}^{(1)}(k\mathbf{r}') \right]. \tag{2.30}$$

This follows trivially from the fact that the VSWFs form a complete, orthogonal basis, and simply states that a VSWF $\widetilde{\mathbf{M}}_{mn}^{(3)}(k\mathbf{r})$ or $\widetilde{\mathbf{N}}_{mn}^{(3)}(k\mathbf{r})$ written in one basis can be expanded in terms of a different basis $\{\widetilde{\mathbf{M}}_{kl}^{(1)}(k\mathbf{r}'), \widetilde{\mathbf{N}}_{kl}^{(1)}(k\mathbf{r}')\}$ with these new eigenfunctions having amplitudes $\{A_{kl}^{mn}, B_{kl}^{mn}\}$. It is easy to write this down; the difficult part is determining the values of A_{kl}^{mn} and B_{kl}^{mn}.

As stated above, the number of elements of the translation matrices is of order n_{max}^4. In the special case of translation along the z axis the matrix elements are only nonzero for $m = k$ [6]:

$$\widetilde{\mathbf{M}}_{mn}^{(3)}(k\mathbf{r}) = \sum_{l=1}^{\infty} \left[A_{ml}^{mn} \widetilde{\mathbf{M}}_{ml}^{(1)}(k\mathbf{r}') + B_{ml}^{mn} \widetilde{\mathbf{N}}_{ml}^{(1)}(k\mathbf{r}') \right],$$

$$\widetilde{\mathbf{N}}_{mn}^{(3)}(k\mathbf{r}) = \sum_{l=1}^{\infty} \left[A_{ml}^{mn} \widetilde{\mathbf{N}}_{ml}^{(1)}(k\mathbf{r}') + B_{ml}^{mn} \widetilde{\mathbf{M}}_{ml}^{(1)}(k\mathbf{r}') \right]. \tag{2.31}$$

Thus an arbitrary translation problem from one sphere to another can be simplified by decomposing the problem into a coordinate rotation, a translation along the z axis and a coordinate rotation back to the initial axes. In the same way as the z translation is orthogonal in n, the rotation is orthogonal in m:

$$\widetilde{\mathbf{M}}_{mn}^{(1)} = e^{im\gamma} \sum_{k=-n}^{n} D_{kn}^{m}(\beta) e^{-k\alpha} \widetilde{\mathbf{M}}_{kn}^{(1)}, \tag{2.32}$$

for Euler angles α, β and γ [9]—and identically for $\widetilde{\mathbf{N}}_{mn}^{(1)}$. Hence:

$$a_{kn} = \sum_{m} D_{kn}^{m} a_{mn}, \tag{2.33}$$

and identically for b_{nm}.

Thus we have decomposed one $O(n_{max}^4)$ transformation into three successive $O(n_{max}^3)$ transformations, which is computationally much faster to calculate.

Expressions for these translation and rotation matrix coefficients were derived and published by Mackowski [9]. However as noted earlier, Mackowski's expressions suffer from combinatorial growth as a function of $(n + m)$, and hence are unsuitable for practical use with spheres many wavelengths in diameter. Consequently we re-derived Mackowski's recurrence relations so they can be applied to our normalized coefficients. The new expressions are derived in Appendix B.

2.6 Forces

2.6.1 Maxwell Stress Tensor

The force on a sphere can be calculated from the Maxwell stress tensor, which represents the net flow of momentum across a surface at a given point. If this is integrated across a surface, the result is the total momentum entering the volume bounded by that surface. In the case of a non-absorbing external medium, we can select a surface which contains exactly one particle within its volume, and the integral will give the total momentum transferred to the sphere. In this way the force on the sphere can be calculated.

The force at a surface element dS can be calculated using Maxwell stress tensor $\overleftrightarrow{T}$, in terms of the electric and magnetic fields $\mathbf{E}$ and $\mathbf{H}$ (and the identity tensor $\overleftrightarrow{I}$) as [13]:

$$dF = \hat{\mathbf{n}} \cdot \overleftrightarrow{T} dS$$

$$\overleftrightarrow{T} = \frac{1}{4\pi}\left(\epsilon\mathbf{EE} + \mathbf{HH} - \frac{1}{2}(\epsilon E^2 + H^2)\overleftrightarrow{I}\right) \tag{2.34}$$

and the net force on the sphere can be determined by integrating over the surface of the sphere:

$$\mathbf{F}_{comp} = \oint_S \hat{\mathbf{n}} \cdot \overleftrightarrow{T} dS. \tag{2.35}$$

While there has been some debate on the exact form of the Maxwell stress tensor, it is accepted [36] that the Minkowski form of the tensor [15, (6.124)] is appropriate for the steady-state case of a solid sphere immersed in a fluid and exposed to optical wavelengths of light - and indeed any such case that is likely to be encountered in optical binding or tweezing contexts, even when deformable droplets are tweezed (see Sect. 2.6.4).

Barton [13] derives an analytical result for the force on a sphere in terms of the VSWF coefficients. That result, re-normalized to suit our normalized VSWF coefficients, is given in Appendix C.1.

2.6.2 Gradient Force

The Maxwell stress tensor (2.34) is used to calculate the *total* optical force on a surface element. While this is convenient, and Barton's analytical result [13] is extremely useful from a computational point of view, it is often helpful from a conceptual point of view to divide the force into the *gradient force* and the *scattering force* [37]. This distinction can be made to a particularly good degree of approximation in the Rayleigh regime; for larger particles this distinction is less clear, although for example the transverse force on a particle in a collimated laser beam is very accurately approximated by the gradient force alone.

There is a simple equation [38] which states the gradient potential U_{grad} (and hence the force $\mathbf{F}_{grad}$) on a particle as a function of the electric field intensity $|E|^2$ contained within the volume of the particle, and the internal and external refractive indices n_{int} and n_{ext}:

$$U_{grad} = -\frac{\epsilon_0}{4}(n_{int} - n_{ext}) \int_V |E|^2 dV \qquad (2.36)$$

$$\mathbf{F}_{grad} = -\nabla U_{grad}$$

This can be evaluated, although being a volume integral it takes a long time to compute. It turns out, though, that the angular part of the integral can be solved analytically, reducing it to a single radial integral. The approach, as in derivations earlier in this chapter, is to substitute the VSWF expansion for the internal field (2.7) into (2.36) and then exploit the orthogonality of the VSWFs (2.42) to solve the integral. After considerable manipulation, we arrive at the following result:

$$U_{grad} = -\frac{\epsilon_0}{4}(n_{int} - n_{ext}) \int \sum_{m,n} \left(|c_{mn}|^2 \mu_n + |d_{mn}|^2 v_n \right) r^2 dr, \qquad (2.37)$$

where μ_n and v_n are the VSWF normalizations defined in (2.42) and c_{mn} and d_{mn} are the expansion coefficients for the field internal to the particle, defined in (2.7). Both these normalizations are functions of r. The radial integral can in fact in turn be solved analytically using the Mathematica symbolic algebra package, but the result is an extremely long combination of hypergeometric functions, and so in practical terms we chose to implement (2.7), complete with single integral, in our code; this single integral can be evaluated quickly enough not to be a significant performance problem.

2.6.3 Absorbing Spheres

Since the Maxwell stress tensor does not make any assumption about non-absorbing media, it should be possible to use the same force formula to calculate the force on absorbing spheres. One concern we encountered in this was the approach taken by Brevik and Sivertsen for a slightly-absorbing sphere [39]. They split the force $\mathbf{F}$ into a surface component $\mathbf{F}_{surf}$ (which was calculated using Barton's formula, treating the sphere as non-absorbing), and a volume absorbing component $\mathbf{F}_{abs}$ which was calculated using a series of intricate integrals. Our concern was that they had *not* used Barton's formula for the absorbing sphere to calculate $\mathbf{F}$ and then subtracted to obtain $\mathbf{F}_{abs} = \mathbf{F} - \mathbf{F}_{surf}$. However in an email

Brevik confirmed [40] that this paper was mostly the work of a student of his, and that he agrees that the Maxwell stress tensor should be applicable in this case.

2.6.4 Forces on Liquid Interfaces

For a liquid droplet it can be useful to calculate the force on an element of the surface of the droplet. This force can be integrated over the surface of the droplet to find the pressure within the droplet, if we assume that the surface tension is sufficient to resist the shear forces on the surface, and maintain the spherical shape of the droplet.

Forces on particles immersed in a dielectric medium have been the source of long-running controversies. The issues are discussed in detail by Brevik [36] and by Nieminen et al. [41]. Many of the problems stem from the question of how momentum is divided between a photon and the polarization of the dielectric medium through which it is propagating. Fortunately this question is largely philosophical in practical cases encountered with optical tweezers setups, and the Minkowski formulation for the Maxwell stress tensor gives accurate results [36]. This is the formulation which Barton used when he derived his formula (our (2.49)).

When calculating the net force on a solid sphere, it is sufficient to determine the net flow of momentum across the surface of the sphere using the Maxwell stress tensor (2.34). For a nonabsorbing sphere, the only parts of the sphere that will experience a force are those parts at the sphere/fluid interface, where there is a step change in the dielectric constant of the medium (resulting in a nonzero $\nabla \epsilon$). If the Maxwell stress tensor is integrated over a surface *within* the volume of the sphere, a force of zero will be found. All momentum transfer occurs at the surface, and within the sphere the momentum of the light field is conserved. If we consider a complete spherical shell outside the surface of the sphere, any net flow of linear momentum into that shell must be transferred to the particle.

In the case of a liquid droplet, however, it is not sufficient to integrate the normal momentum flow $\hat{\mathbf{n}} \cdot d\mathbf{p} = \hat{\mathbf{n}} \cdot (\hat{\mathbf{n}} \cdot \overleftrightarrow{T}) dS$ over a surface just outside the sphere. For a given surface element we must look at the difference between the momentum density of the field just *outside* the surface, and of the field just *inside* the surface. The difference between their normals gives the total force compressing (or expanding) that element of surface. A naive calculation which integrates the net flow of momentum across a surface S' just outside the sphere will overestimate the force on the interface as it will have calculated the *entire momentum flow* across that surface, rather than that fraction of the momentum which is transferred to the dielectric interface (which can be deduced by looking at the *difference* between the flow across S and the flow across a surface S' just *inside* the sphere). We must therefore evaluate that difference in order to find the instantaneous compressive force $\mathbf{F}_{comp}$:

$$\mathbf{F}_{comp} = \oint_{S} \hat{\mathbf{n}} \cdot (\hat{\mathbf{n}} \cdot \overleftrightarrow{T})dS - \oint_{S'} \hat{\mathbf{n}} \cdot (\hat{\mathbf{n}} \cdot \overleftrightarrow{T})dS' \qquad (2.38)$$

where S' is a surface just inside the sphere and S a surface just outside the sphere.

We can use this integral to determine the average pressure on the surface of the droplet. If we assume that the droplet is only slightly deformed before the surface tension associated with the deformation creates an equilibrium in the surface forces, then the pressure inside the droplet is simply the mean value of this integral over the surface. We were in fact able to solve this integral analytically, and the derivation is given in Appendix C.2. This pressure can for example be used to determine the direction and rate of flow between two droplets which are joined by a narrow "thread" of liquid.

Alternatively, the *distribution* of forces across the surface of the sphere can be exploited in conjunction with a surface tension model to calculate the deformation of the sphere as a result of these forces. This is something that we have been investigating in collaboration with Colin Bain (Department of Chemistry, Durham) and Alex Lubansky (Department of Engineering, Oxford) [42, 43].

2.6.4.1 Electrostriction

One factor that deserves specific discussion in this context is *electrostriction*, since this effect will occur in the case of two liquid droplets joined by a thin tube, which is a configuration we are interested in. This issue is not explicitly discussed in the literature as far as we are aware.

Electrostriction is in a sense an extension of the gradient force effect (Sect. 2.6.2) to deformable media. The gradient potential follows from the fact that a dielectric object's energy is lowered if it is situated in a region of higher electric field strength. The gradient potential of a deformable object is a function not only of its position but of its shape, and its energy may be lowered if it is deformed to better fit the electric field distribution around it, increasing the integrated intensity contained within its volume. This results in an *electrostrictive* force on the surface of the object. For example, an incompressible liquid droplet will minimize its energy in a laser beam by elongating itself along the beam axis.[3] The electrostrictive force is given by the following equation [36]:

---

[3] Similar forces occur on solid objects, but they are balanced out by mechanical stresses in the object [36].

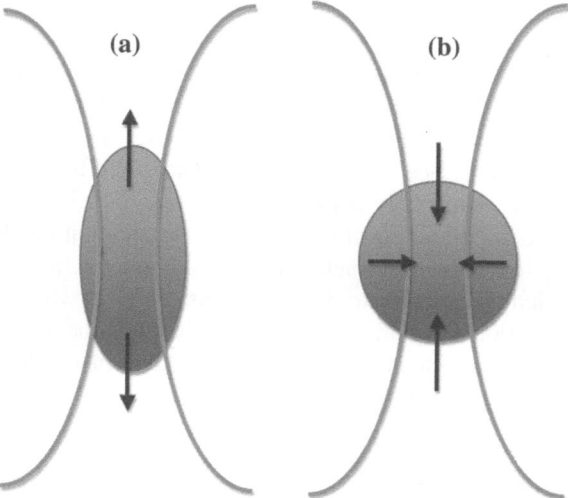

Fig. 2.5 Forces (black arrows) acting on a deformable oil droplet (blue) in optical tweezers (beam represented schematically in yellow). **a** Minkowski force (as described by the *Maxwell stress tensor*): the force acts on a dielectric surface where there is a nonzero electromagnetic field, pulling the surface out. This results in deformation of the droplet. **b** Electrostrictive force (2.39): the force acts on surface and at edges of beam within the oil. This results in an increased oil pressure along the polar axis of the droplet relative to the ressure around its equator, but this should not have any effect on the shape of the droplet surface, since this increased pressure balances the electrostrictive force on the surface. Note though that for a *compressible* droplet, electrostriction would slightly decrease the overall volume of the droplet

$$\mathbf{F} = -\frac{1}{2}\nabla\left[E^2\rho_m\left(\frac{\delta\epsilon}{\delta\rho_m}\right)_T\right]$$
$$\left(\frac{\delta\epsilon}{\delta\rho_m}\right)_T = \frac{(\epsilon-1)(\epsilon+2)}{3\rho_m} \tag{2.39}$$

where ρ_m is the mass density of the medium and the partial derivative is the Clausius-Mossotti relation.

As a result of this force, pressures will be greater within the region of liquid contained within the beam (as illustrated in Fig. 2.5). However the pressure in the connecting tube, outside this region, will be unaffected by electrostriction, and so electrostriction will have no effect on the flow of liquid between the two droplets.

The electrostrictive effect will produce forces at the surface of a droplet which are of the same order, or larger, than the forces we have used in our calculations of the surface pressure using the Maxwell stress tensor. However, the electrostrictive forces arise from a conservative potential, and for an incompressible droplet they will have no effect on the droplet shape, even in the case where there is a fluid outside the droplet (rather than a vacuum). Figure 2.5 helps to explain the reasoning behind this. Although an additional (inward) pressure will be present on *some regions* of the droplet surface, that force is exactly balanced by an increase in the internal pressure of the equivalent regions inside the droplet. The result of the electrostrictive effect is that the actual pressure within the droplet is *not uniform* even when the droplet surface is in equilibrium. There will be a higher pressure in regions where the laser intensity is higher.

There will therefore be no effect due to electrostriction on the droplet shape or on the flow of liquid from one droplet to another. Although a full description of the forces should include the electrostrictive effect, its omission will not introduce any error in the deformation calculation. There would in theory be a slight modification to the reaction rate of any chemical reaction which takes place within the droplets, but for liquid reagents this alteration would be negligible.

2.7 Motion

One very important application for the forces on a particular configuration of particles is the calculation of the resultant particle trajectories over time. The simplest approach is the *Euler method* which for a velocity $\mathbf{v}$ at time t calculates the position $\mathbf{r}'$ at time $t + \Delta t$ as $\mathbf{r}' = \mathbf{r} + \mathbf{v}\Delta t$. More efficient and accurate methods such as the Runge-Kutta method exist [44, Sect. 16.1], but these are intended for use with continuously differentiable functions, which makes them poorly suited to treating random Brownian motion. We treat the system as over-damped, and calculate the instantaneous velocity $\mathbf{v}$ using Stokes' law: $\mathbf{v} = \mathbf{f}/D_s$ where the Stokes drag coefficient D_s for a sphere of radius a is $D_s = 6 \times 10^{-3}\pi a$. The instantaneous force $\mathbf{f}$ is calculated by combining the force due to the electromagnetic field with any forces due to electrostatic charges, hydrodynamic interactions, gravity and Brownian motion (as required).

We will not discuss the details of the motion calculations here, since we have used recognised techniques which are widely discussed in the relevant literature [45–48]. We will simply summarize the two methods we have used used.

The first method is for the treatment of Brownian motion. In some cases, particularly for particles smaller than the wavelength of light and in the presence of interference fringes [49, 50], the particles are moving in an optical "landscape" consisting of a large number of weak potential minima. A model which does not take Brownian motion into account will be unable to describe the evolution of such a system, because the simulated particles will simply become trapped in a local minimum from which they will never escape, despite the fact that they would in reality have enough thermal energy to escape from that potential well. In order to simulate this situation we use the Euler method. As the particles move under the influence of Brownian motion, the forces on them due to their interaction with the light field will change. Qualitatively, once any one particle has moved a distance which is at all comparable with the wavelength of light then the light force on all the particles in the system is liable to change significantly. Consequently, very short timesteps Δt must be used in order to meet this constraint and ensure physically realistic simulations. For simulations of particles of the order of 500 µm in diameter, it turns out that a timestep $\Delta t \sim 10\,\mu m$ must be used, which makes such simulations very time-consuming.

The second method ignores Brownian motion, and uses a 4th-order Runge-Kutta method with 5th-order error control to integrate the equation of motion of the particle. This method is not ideal, since the system is a fairly stiff one, but this proved to be the best-performing of the commonly-available techniques that we evaluated.

In both cases, it is possible to approximate the hydrodynamic interactions between multiple nearby particles. This effect can be significant when the particles are within a few radii of each other [45, 46, 48]. This is achieved by multiplying the vector containing the particle velocities by a correction matrix known as the *Oseen tensor* [46]. This same matrix can also be used to take into account the correlations which will occur between the random Brownian motion of nearby particles [47].

One positive aspect of the fact that multiple-particle Mie scattering models are computationally hard to solve is that this means quite sophisticated additional calculations can be performed at each timestep without significant affecting the overall speed of the simulation. This means that these hydrodynamic and electrostatic effects can be easily included in the model.

Appendix A: Special Functions and Physical Quantities

A.1 Special Functions

P_n^m	Associated Legendre polynomial
$\widetilde{\mathbf{M}}_{mn}^{(j)}(r,\theta,\phi)$ $\widetilde{\mathbf{N}}_{mn}^{(j)}(r,\theta,\phi)$	Normalized vector spherical wavefunctions — defined below in (2.40)
$\widetilde{\pi}_{mn}(\cos\theta)$ $\widetilde{\tau}_{mn}(\cos\theta)$	Un-named normalized angular functions — defined below in (2.41)
J_n	Bessel function of the first kind, of order n
j_n	Spherical Bessel function of the first kind, of order n
n_n	Spherical Bessel function of the second kind, of order n
$h_n^{(1)} = j_n + i\,n_n$	Spherical Hankel function of the first kind

A.2 Physical Quantities and Other Symbols Used

ϵ_{ext}	Permittivity of external medium
$n_{ext} = \sqrt{\epsilon_{ext}}$	Refractive index of external medium
n_s	Refractive index of a sphere
k	Wave number of the laser in the external medium
a	Radius of a sphere
w_0	Gaussian beam waist radius
$\mathbf{E}$	Electric field vector
$I = |\mathbf{E}|^2$	Electric field intensity
$\mathbf{H}$	Magnetic field vector

A.3 Vector Spherical Wavefunctions

The vector spherical wavefunctions $\widetilde{\mathbf{M}}_{mn}^{(j)}$ and $\widetilde{\mathbf{N}}_{mn}^{(j)}$ are defined as follows:

$$\widetilde{\mathbf{M}}_{mn}^{(j)}(r,\theta,\phi) = \left[i\widetilde{\pi}_{mn}(\cos\theta)\mathbf{i}_\theta - \widetilde{\tau}_{mn}(\cos\theta)\mathbf{i}_\phi \right] e^{im\phi} z_n^{(j)}(kr),$$

$$\widetilde{\mathbf{N}}_{mn}^{(j)}(r,\theta,\phi) = \frac{n(n+1)}{kr}\sqrt{\frac{2n+1}{4\pi}\frac{(n-m)!}{(n+m)!}} P_n^m(\cos\theta) e^{im\phi} z_n^{(j)}(kr)\mathbf{i}_r$$

$$+ \left[\widetilde{\tau}_{mn}(\cos\theta)\mathbf{i}_\theta + i\widetilde{\pi}_{mn}(\cos\theta)\mathbf{i}_\phi \right] e^{im\phi} \frac{1}{kr}\frac{\mathrm{d}}{\mathrm{d}(kr)}\left((kr)z_n^{(j)}(kr) \right),$$

$$(2.40)$$

where

$$\widetilde{\pi}_{mn}(\cos\theta) = \sqrt{\frac{2n+1}{4\pi}\frac{(n-m)!}{(n+m)!}}\frac{m}{\sin\theta}P_n^m(\cos\theta),$$

$$\widetilde{\tau}_{mn}(\cos\theta) = \sqrt{\frac{2n+1}{4\pi}\frac{(n-m)!}{(n+m)!}}\frac{\mathrm{d}}{\mathrm{d}\theta}P_n^m(\cos\theta), \qquad (2.41)$$

$$z_n^{(1)}(kr) = j_n(kr),$$

$$z_n^{(3)}(kr) = h_n^{(1)}(kr).$$

They have the property that they are mutually orthogonal when integrated over a spherical surface:

$$\mu_n = \int_S \widetilde{\mathbf{M}}_{mn}^{(j)} \widetilde{\mathbf{M}}_{m'n'}^{(j)} dS = n(n+1)|z_n|^2 \delta_{m,m'}\delta_{n,n'},$$

$$v_n = \int_S \widetilde{\mathbf{N}}^{(j)}_{mn} \widetilde{\mathbf{M}}^{(j)}_{m'n'} dS$$

$$= \left([n(n+1)]^2 \left| \frac{z_n}{kr} \right|^2 + n(n+1) \left| \frac{1}{kr d(kr)} (kr\, z_n) \right| \right) \delta_{m,m'} \delta_{n,n'}, \int_S \widetilde{\mathbf{M}}^{(j)}_{mn} \widetilde{\mathbf{N}}^{(j)}_{m'n'} dS = 0.$$

$$(2.42)$$

Appendix B: Generation of Translation and Rotation Matrix Coefficients

This appendix derives expressions for calculating the translation and rotation matrix coefficients A^{mn}_{ml}, B^{mn}_{ml} and D^m_{kn} which form the components of the matrix **F** in (2.4), which is used in transforming from a VSWF basis centred on one position to one centred on a different position.

Equivalents to these expressions were derived and published by Mackowski [9]. However as noted earlier, Mackowski's expressions suffer from combinatorial growth as a function of $(n + m)$, and hence are unsuitable for practical use with large spheres. Consequently we have modified Mackowski's recurrence relations so they can be applied to our normalized coefficients. We emphasize that the original derivation of the equations given below was carried out by Mackowski, but the derivation is repeated here with different normalization of the coefficients. In addition, a number of minor errors in Mackowski's expressions are corrected.

To generate the rotation coefficients $D^m_{kn}(\alpha, \beta, \gamma)$ we start with the analytical result [9, Equation 89]:

$$D^0_{kn} = \sqrt{\frac{(n+k)!}{(n-k)!}} P^{-k}_n (\cos \beta), \qquad (2.43)$$

and then for a given k, n we can use the following recurrences to generate the values for all m:

$$D^{m+1}_{kn} = \frac{1}{\sqrt{(n+m+1)(n-m)}} \left[\alpha \sqrt{(n+k)(n-k+1)} D^m_{k-1,n} \right.$$

$$\left. - \beta \sqrt{(n-k)(n+k-1)} D^m_{k+1,n} - k\gamma D^m_{kn} \right],$$

$$D^{m-1}_{kn} = \frac{1}{\sqrt{(n-m+1)(n+m)}} \left[-\alpha \sqrt{(n+k)(n-k+1)} D^m_{k-1,n} \right.$$

$$\left. + \beta \sqrt{(n-k)(n+k+1)} D^m_{k+1,n} - k\gamma D^m_{kn} \right]. \qquad (2.44)$$

The translation coefficients A_{ml}^{mn} and B_{ml}^{mn} are generated from the scalar translation coefficients C_{ml}^{mn} as follows:

$$A_{m,l}^{m,n} = kz\sqrt{\frac{(l+m+1)(l-m+1)}{(l+1)^2(2l+1)(2l+3)}} C_{m,l+1}^{m,n}$$
$$+ kz\sqrt{\frac{(l+m)(l-m)}{(l)^2(2l-1)(2l+1)}} C_{m,l-1}^{m,n} \qquad (2.45)$$
$$+ C_{m,l}^{m,n},$$
$$B_{m,l}^{m,n} = \frac{imkz}{l(l+1)} C_{m,l}^{m,n}.$$

We start with the following seed values for the scalar translation coefficients:

$$C_{0,l}^{0,0} = (2l+1)h_l(kz),$$
$$C_{0,0}^{0,n} = h_n(kz). \qquad (2.46)$$

The following recurrences then yield the values for $m = n$ and finally for arbitrary m:

$$C_{\pm n,l}^{\pm n,n} = \sqrt{\frac{(2n+1)(l+n-1)(l+n)}{(2l+1)(2n)(2l-1)}} C_{\pm(n-1),l-1}^{\pm(n-1),(n-1)},$$
$$+ \sqrt{\frac{(2n+1)(l-n+1)(l-n+2)}{(2l+1)(2n)(2l+3)}} C_{\pm(n-1),l+1}^{\pm(n-1),(n-1)},$$
$$C_{m,l}^{m,n} = \sqrt{\frac{(2n+1)(n-m-1)(n+m-1)}{(2n-3)(n+m)(n-m)}} C_{m,l}^{m,n-2} \qquad (2.47)$$
$$+ \sqrt{\frac{(2n+1)(2n-1)(l+m)(l-m)}{(2l+1)(2l-1)(n+m)(n-m)}} C_{m,l-1}^{m,n-1}$$
$$- \sqrt{\frac{(2n+1)(2n-1)(l+m+1)(l-m+1)}{(2l+1)(2l+3)(n+m)(n-m)}} C_{m,l+1}^{m,n-1}.$$

As Mackowski points out, the recurrences for C_{ml}^{mn} do not contain any reference to the translation distance kz. The recurrences are entirely geometrical relationships, and it is purely the initial seed values that determine the distance of the translation.

The following recurrences can also be useful in reducing the amount of work required in generating the coefficients:

$$C_{-m,n}^{-m,l} = (-1)^{l+n} C_{m,l}^{m,n},$$
$$D_{-k,n}^{-m} = (-1)^{k+m} D_{k,n}^{m}. \qquad (2.48)$$

Appendix C: Forces on a Particle

C.1 Net Force on a Particle

Barton [13] derived an analytical result for the force on a sphere (Eq. 2.35) in terms of the VSWF coefficients. We state that result below, after re-normalizing it to suit our normalized VSWF coefficients.

$$
\begin{aligned}
F_x + iF_y = & -il(l+2)\sqrt{\frac{(l+m+2)(l+m+1)}{(2l+1)(2l+3)}} \\
& \times (a_{nm}p^*_{n+1,m+1} + b_{nm}q^*_{n+1,m+1} \\
& \quad + p_{nm}a^*_{n+1,m+1} + q_{nm}b^*_{n+1,m+1} \\
& \quad - 2a_{nm}a^*_{n+1,m+1} - 2b_{nm}b^*_{n+1,m+1}) \\
& -il(l+2)\sqrt{\frac{(l-m+1)(l-m+2)}{(2l+1)(2l+3)}} \\
& \times (a_{n+1,m-1}p^*_{nm} + b_{n+1,m-1}q^*_{nm} \\
& \quad + p_{n+1,m-1}a^*_{nm} + q_{n+1,m-1}b^*_{nm} \\
& \quad - 2a_{n+1,m-1}a^*_{nm} - 2b_{n+1,m-1}b^*_{nm}) \\
& + \sqrt{(l+m+1)*(l-m)} \\
& \times (a_{nm}q^*_{n,m+1} + b_{nm}p^*_{n,m+1} \\
& \quad + q_{nm}a^*_{n,m+1} + p_{nm}b^*_{n,m+1} \\
& \quad - 2a_{nm}b^*_{n,m+1} - 2b_{nm}a^*_{n,m+1}),
\end{aligned}
$$

(2.49)

$$
\begin{aligned}
F_z = \mathrm{Im}\Bigg[& l(l+2)\sqrt{\frac{(l-m+1)(l+m+1)}{(2l+3)(2l+1)}} \\
& \times (a_{n+1,m}p_{nm} + b_{n+1,m}q_{nm} \\
& \quad + p_{n+1,m}a_{nm} + q_{n+1,m}b_{nm} \\
& \quad - 2a_{n+1,m}a_{nm} - 2b_{n+1,m}b_{nm}) \\
& - im(a_{nm}q_{nm} + b_{nm}p_{nm} \\
& \quad + p_{nm}b_{nm} + q_{nm}a_{nm} \\
& \quad - 2a_{nm}b_{nm} - 2b_{nm}a_{nm})\Bigg].
\end{aligned}
$$

(2.50)

C.2 Pressure Inside a Liquid Droplet

Equation 2.38 gives the pressure inside a spherical liquid droplet exposed to a laser beam. This appendix gives our derivation for the analytical solution to this integral in terms of the VSWF coefficients a, b, c, d, p and q defined in Sect. 2.3 (for brevity we will condense a_{mn} to a, b_{mn} to b etc.).

We start by expanding the integral in (2.38) for the normal momentum flux just outside and inside the droplet surface, (using the *peak* amplitudes of the fields, E and H):

$$F = \frac{1}{16\pi} \int_S \left(2\epsilon E_r E_r^* - \epsilon|E|^2 + 2H_r H_r^* - |H|^2 \right) \sin\theta d\theta d\phi. \qquad (2.51)$$

Substituting in the expansions for the incident, scattered and internal fields we have:

$$\frac{16\pi F_{ext}}{\epsilon_{ext}} = 2\sum_{m,n} \int_S \left(\left| aN_r^{(3)} - pN_r^{(1)} \right|^2 \right) dS$$

$$+ 2\sum_{m,n} \int_S \left(\left| bN_r^{(3)} - qN_r^{(1)} \right|^2 \right) dS$$

$$- \sum_{m,n} \int_S \left(\left| a\mathbf{N}^{(3)} - b\mathbf{M}^{(3)} - p\mathbf{N}^{(1)} - q\mathbf{M}^{(1)} \right|^2 \right) dS$$

$$- \sum_{m,n} \int_S \left(\left| b\mathbf{N}^{(3)} - a\mathbf{M}^{(3)} - q\mathbf{N}^{(1)} - p\mathbf{M}^{(1)} \right|^2 \right) dS \qquad (2.52)$$

$$\frac{16\pi F_{int}}{\epsilon_{ext}} = 2\sum_{m,n} \int_S \left| -dN_r^{(1)} \right|^2 dS$$

$$+ 2\sum_{m,n} \int_S \left| -cN_r^{(1)} \right|^2 dS$$

$$- \sum_{m,n} \int_S \left| -d\mathbf{N}^{(1)} - c\mathbf{M}^{(1)} \right|^2 dS$$

$$- \sum_{m,n} \int_S \left| -c\mathbf{N}^{(1)} - d\mathbf{M}^{(1)} \right|^2 dS$$

Exploiting the orthogonality and normalization of the VSWFs (2.42), we can solve these integrals to obtain:

$$F = F_{ext} - F_{int}$$

$$F_{ext} = \sum_{m,n} A^{33}(aa^* + bb^*) + A^{11}(pp^* + qq^*) + A^{13}(ap^* + pa^* + bq^* + qb^*)$$

$$F_{int} = \sum_{m,n} B^{33}(cc^* + dd^*),$$ (2.53)

where:

$$A^{ij}, B^{ij} = \left(\left(\frac{n(n+1)}{\alpha} \right)^2 - n(n+1) \right) z^{(i)}(\alpha) z^{(j)}(\alpha)$$

$$- \frac{n(n+1)}{(\alpha)^2} \frac{d}{d\alpha} \left(\alpha z^{(i)}(\alpha) \right) \frac{d}{d\alpha} \left(krz^{(j)}(\alpha) \right)$$ (2.54)

$$(\alpha = kr \text{ for } A^{ij}, \alpha = k'r \text{ for } B^{ij},$$

where k' is the wavenumber inside the sphere).

Thus we have eliminated the surface integrals, and the force is expressed in terms of products of the field coefficients a, b etc. with parameters A^{ij}, B^{ij} that depend only on the physical properties of the sphere.

Appendix D: Implementation and Optimization

The schematic process for calculating the field and force for a particular particle and beam configuration is shown in Fig. 2.6. This can then be used as the building block within a dynamics simulation which models the evolution of that system over time (Fig. 2.7).

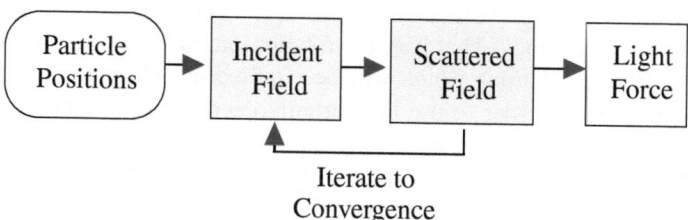

Fig. 2.6 Flow diagram showing the procedure for determining the field and forces for a given particle configuration. For the specified positions, the calculation is seeded with the incident field from the laser, and the scattering for that field is calculated. This scattering is then included in the incident field in the next round of the iteration. When the iterative calculation has converged, the force is calculated

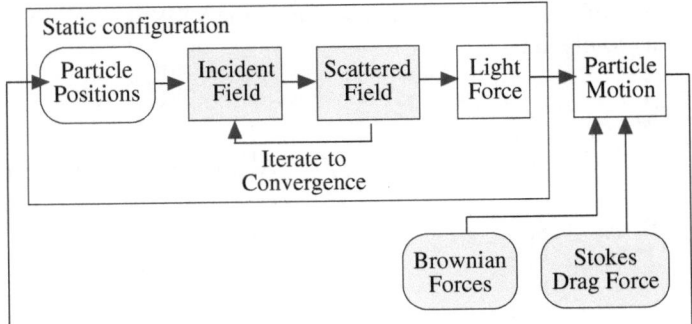

Fig. 2.7 Flow diagram showing the simulation of particle motion as a function of time. The force is calculated for a static configuration, as shown in Fig. 2.6, and then the force for that configuration is used to determine the instantaneous motion of the particles, taking into account such effects as Brownian motion, the Stokes drag force, and hydrodynamic interactions between nearby particles if required

We implemented these algorithms from scratch in computer code. This was a very considerable task, with the main challenges being the correct implementation of all the many mathematical equations involved, verification that the output of the code is correct, and optimization of the code to ensure that it runs fast enough to be able to perform time evolution simulations within a feasible computation time. This issue of performance is an important one: a major obstacle to being able to perform time evolution simulations of Mie scattering models is the computing power required to carry out the calculations.

D.1 Geometric Considerations

The rotational/translational decomposition of the translation matrices (described in Sect. 2.5) means that an important factor determining the speed a problem can be solved is the geometry of the problem. Selection of the right cartesian axes for the problem can speed it up considerably.

In the case of a chain of particles (constrained to a 1D line) the particles should be orientated along the z axis. That way the rotation matrices can be dispensed with entirely—all the translations which will be required are along the z axis and no rotation is required in order to use the partially decoupled formulation in (2.31). With the axes fixed in this way by the orientation of the particles, the propagation vector of the laser beam (defined relative to the particle orientation) is also fixed. The formula for the beam coefficients may simply permit this propagation vector to be used directly (e.g. for a plane wave, where (2.12) accepts an arbitrary propagation vector **k**). In other cases, such as the Bessel beam formula in Sect. 2.4.4, the formula does not easily lend itself to a change in orientation. It will nevertheless be considerably more efficient to apply an appropriate rotation matrix to n spheres' external fields, to ensure the required orientation of the beam in the chosen coordinate system, than it is to apply a rotation to $n^2/2$ translation operations on *every* iteration of the field.

This advantage of decoupling the VSWF coefficients of differing m is also of enormous importance when solving the field by inversion (Sect. 2.5.2, [6]). This allows the one huge equation system to be split into $2n_{max} + 1$ independent smaller systems, vastly reducing the computation time and memory requirements for the algorithm.

In the case of particles constrained to a 2D surface (as may be specified when modelling evanescent wave experiments), it is desirable to have the surface lie in the x/y plane. In that case, every translation will include a rotation, but it will always be a rotation by exactly 90°. Only that single rotation matrix needs to be calculated, and the iterative process can also be considerably quicker: the presence of data caches in modern computers mean that accessing a single block of memory $n^2/2$ times is considerably quicker than accessing $n^2/2$ separate blocks of memory.

D.2 Code Performance

The code is structured in a hierarchical manner which ensures very fast code for the time-critical calculations (beam expansion, translation and force calculation) while minimizing the programming effort required.

The lowest-level building block is based around a unit consisting of a pair of complex numbers, which could for example represent the pair of VSWF coefficients $\{a_{mn}, b_{mn}\}$. This is coded in the C++ language as a *template*, which defines primitive arithmetic operations, such as addition, between these objects in terms of highly efficient sequences of SSE3 assembly language instructions.[4] Once this single object has been defined then all the code involving these complex number pairs can be written in a high-level language, and the compiler will automatically translate the code into this very fast underlying implementation.

The next level in the hierarchy is the implementation of the various linear algebra equations representing operations such as translations. As can be seen from Sects. 2.2 and 2.5, these fundamental calculations can be represented as a series of matrix operations. There are off-the-shelf packages for most computer platforms and languages which can efficiently perform linear algebra calculations. However the matrices involved in our case are sparse (although there are well-defined rules which define which matrix elements are empty and which are not). One example of this is the condition on the indices k, m and n requiring $|k| \leq n$ as well as $|m| \leq n$. Unfortunately, this means that, if the code is to perform effectively, then no off-the-shelf solution is appropriate, and appropriate code must be written by hand. However once these building blocks have been implemented, the higher-level code, which is not performance critical, can in principle be written in any desired

---

[4] Streaming SIMD Extensions 3, allowing several identical calculations to be carried out simultaneously by the processor.

Fig. 2.8 Performance comparison on various computing platforms (*shorter is better*). The categories indicate the number of particles in the configuration; "sparse" refers to a configuration with a chain of particles 10 radii apart, and "dense" to a 2D arrangement of particles around 2.5 radii apart; "micro" refers to micron-sized particles ($n_{max} = 45$) and "nano" to sub-micron particles ($n_{max} = 12$). **a** shows the time taken to generate the required translation matrices and **b** the time taken for the iterative field calculation to converge. A general trend can be seen where the Mac Pro is (unsurprisingly) the fastest.[6]

(a)   Time taken for matrix generation

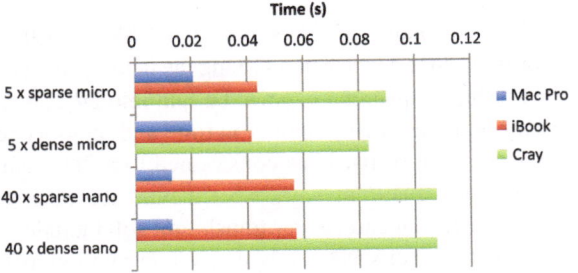

(b)   Time taken for field iteration

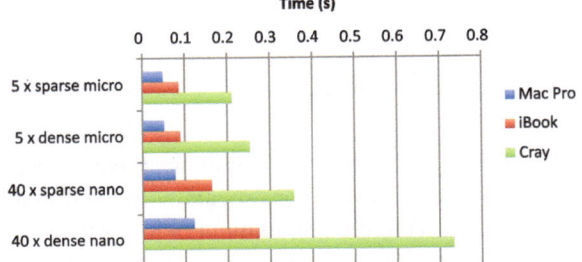

language. There is the option of choosing a different language which prioritizes clarity and conciseness over performance considerations.

The highest level in the hierarchy is scheduling code which is designed to divide the work up between multiple processors (on the level of individual matrix operations or integrations), and even multiple computers (on the level of separate field calculations[5]).

---

[5] The considerable quantity of data which must be exchanged between particles in a single field calculation generally precludes dividing an individual field calculation up between multiple computers, although such a calculation scales well across multiple processors in a single computer

[6] It is interesting to note that in contrast to the other platforms the Mac Pro takes *less time* to calculate the matrices for 40 small particles than it does for 5 large particles, which is indicative that the large number of small calculations scales well to the 8 available processor cores, whereas the scaling is less good for a smaller number of larger calculations. Recent advances in operating system technologies have considerable potential for addressing this issue if a little time could be devoted to updating our computer code [53].

D.3 Computer Platforms Used

A number of different computer platforms were investigated and used for the computer code. The main ones used were the following:

- Apple Macintosh iBook (2.2 GHz Core 2 Duo). This was the development system, and was also easily capable of running moderate-length sequences of field calculations.
- Cray XD1 (2 × 2.1 GHz AMD Opteron, 6 nodes). This resource was available within the research group, and so quite a few simulations were run on it, but its performance has since been surpassed by more modern platforms.
- Apple Macintosh Mac Pro (8 × 3 GHz Intel Xeon). This was invaluable for long dynamics simulations, as demonstrated in the performance comparison given later.
- Sony Playstation 3 (PS3). This low-cost system based on the Cell Broadband Engine is remarkably powerful, but requires considerable platform-specific expertise and coding effort, as well as suffering from memory limitations. The memory limitations mean that it is only suited to problems with a small value of n_{max} (without considerable additional programming effort), but for problems such as that in Chap. 3 it excels. This is discussed in more detail below.

Figure 2.8a–c summarise the relative performance of the different computing platforms used. As described in the figure, the Mac Pro considerably out-performs the iBook and Cray platforms, and the code I have written is slightly more efficient for problems with larger numbers of spheres on the Mac Pro platform. As Fig. 2.8c shows, the PS3 is actually about 50% *faster* than the Mac Pro when dealing with a dense configuration of 40 small spheres, despite being one fifth of the purchase cost. This advantage could probably be further improved if more time was invested in improving the platform-specific code required by the PS3.

For some applications, then, it would be an excellent investment to buy a number of PS3s, but since there is sufficient general-purpose computing resource available within the group, we did not tend to use the PS3 a great deal since it does require some time to be expended in hand-tailoring the code to the platform.

Finally, Fig.2.8d compares our code to the Fortran code written by Yu-Lin Xu [51, 52], which to our knowledge is the only publicly-available code with capabilities comparable to ours. It can be seen than even when our own code is limited to a single thread of execution for the purposes of the comparison,[7] our own code is 35 and 70 times faster than Xu's for the two examples shown.

---

[7] This causes our code to run at reduced speed, but provides a fairer comparison since the Fortran code can only make use of a single processor core; multiple cores can best be exploited by running several completely independent calculations in parallel.

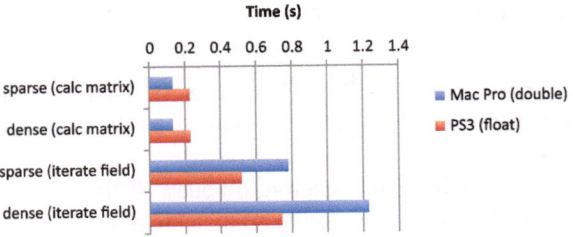

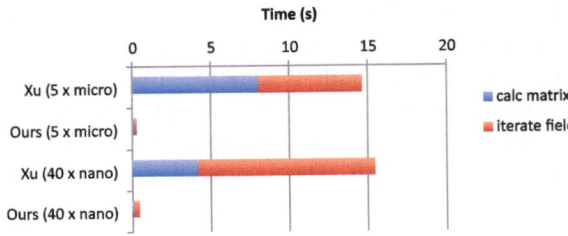

Fig. 2.9 Performance comparison on two different computing platforms, and comparison with existing publicly-available code (shorter is better). The categories indicate the number of particles in the configuration; "sparse" refers to a configuration with a chain of particles 10 radii apart, and "dense" to a 2D arrangement of particles around 2.5 radii apart; "micro" refers to micron-sized particles ($n_{max} = 45$) and "nano" to sub-micron particles ($n_{max} = 12$). **a** compares the performance of the PS3 and the Mac Pro for $n_{max} = 12$ (the PS3 code is unable to handle the larger spheres). Note also that the PS3 only performs a single-precision (reduced accuracy) calculation. The comparison shows that the PS3 is *faster* than the Mac Pro despite its considerably lower specification (and cost). **b** compares the performance of our code to that of Xu's Fortran code [51, 52] for two of the cases shown in Fig. 2.8. Our code can be seen to be around 70 times faster for the case of 5 large spheres, and around 35 times faster for the case of 40 small spheres

D.4 Testing

Because of the scale of the computer code developed for the Mie calculations, it was important to be able to verify the correctness of its results, and to have a self-testing suite which can be run at regular intervals to check that core function has not accidentally been broken by enhancements to the code. The self-test process tested the building blocks of the calculation against more general (but slower) calculations, verified that conditions such as the boundary conditions on dielectric boundaries were satisfied, and that consistent results are output when an identical problem is formulated in different orientations relative to the coordinate axes. It also verified results against a range of published results in the literature [13, 54–56].

In the literature there was a significant gap in terms of quantitative results for counter-propagating beam traps in water, which was a considerable cause for concern since this is one of the main configurations we are interested in. In an attempt to address that issue, we implemented a primitive finite difference time domain (FDTD) simulation [57]. Although extremely slow, this method has the advantage of being entirely independent from Mie scattering from a methodological point of view, and therefore provides a very good source of data for comparison. In recent years, a number of groups have now published results from Mie scattering and related calculations [12, 45, 58], which our code agrees with.

References

1. Osborn, J.: Private communication. (2008)
2. Mie, G.: Contribution to the optical properties of turbid media, in particular of colloidal suspensions of metals. Ann. Phys. **25**, 377 (1908)
3. Debye, P.: Der lightdruck auf kugeln von beliebigem material. Ann. Phys. **30**, 57–136 (1909)
4. Stratton, J.A.: Electromagnetic Theory. McGraw-Hill, New York (1941)
5. Liang, C., Lo, Y.T.: Scattering by two spheres. Radio Sci. **2**, 1481–1495 (1967)
6. Bruning, J.H., Lo, Y.T.: Multiple scattering of waves by spheres. IEEE Trans. Antennas Propag. **AP-19**, 378–400 (1971)
7. Fuller, K.A., Kattawar, G.W.: Consummate solution to the problem of classical electromagnetic scattering by an ensemble of spheres 1: linear chains. Opt. Lett. **13**(2), 90–92 (1988)
8. Fuller, K.A., Kattawar, G.W.: Consummate solution to the problem of classical electromagnetic scattering by an ensemble of spheres 2: clusters of arbitrary configuration. Opt. Lett. **13**(12), 1063–1065 (1988)
9. Mackowski, D.W.: Analysis of radiative scattering for multiple sphere configurations. Proc. R. Soc. London, Ser. A **433**, 599–614 (1991)
10. Bohren, C.F., Huffman, D.R.: Absorption and Scattering of Light by Small Particles. Wiley, New York (1983)
11. Barton, J.P., Alexander, D.R., Schaub, S.A.: Internal and near-surface electromagnetic fields for a spherical particle irradiated by a focused laser beam. J. Appl. Phys. **64**(4), 1632–1639 (1988)
12. Čižmár, T., Kollárová, T., Bouchal Z., Zemánek, P.: Sub-micron particle organization by self-imaging of non-diffracting beams. New J. Phys. **8**, 43 (2006)
13. Barton, J.P., Alexander, D.R., Schaub, S.A.: Theoretical determination of net radiation force and torque for a spherical particle illuminated by a focused laser beam. J. Appl. Phys **66**(10), 4594–4602 (1989)
14. Xu, Y.-L.: Electromagnetic scattering by an aggregate of spheres. Appl. Opt. **34**(21), 4573–4588 (1995)
15. Jackson, J.D.: Classical Electrodynamics. Wiley, New York (1975)
16. van de Hulst, H.C.: Light Scattering by Small Particles. Dover, New York (1981)
17. Kattawar, G.W., Plass, G.N.: Electromagnetic scattering from absorbing spheres. Appl. Opt. **6**(8), 1377–1382 (1967)
18. Wiscombe, W.J.: Improved Mie scattering algorithms. Appl. Opt. **19**, 1505–1509 (1980)
19. Mackowski, D.W.: Calculation of total cross sections of multiple-sphere clusters. J. Opt. Soc. Am. A **11**(11), 2851–2861 (1994)
20. Chew, H., Wang, D.-S., Kerker, M.: Elastic scattering of evanescent electromagnetic waves. Appl. Opt. **18**(15), 2679–2687 (1979)

21. Garcés-Chávez, V., McGloin, D., Melville, H., Sibbett, W., Dholakia, K.: Simultaneous micromanipulation in multiple planes using a self-reconstructing light beam. Nature **419**, 145–147 (2002)
22. Garcés-Chávez, V., Roskey, D., Summers, M.D., Melville, H., McGloin, D., Wright, E.M., Dholakia, K.: Optical levitation in a bessel light beam. Appl. Phys. Lett. **85**, 4001–4003 (2004)
23. Abramowitz, M., Stegun, I.A.:Handbook of Mathematical Functions. Dover, New York (1972)
24. Khaled, E.E.M., Hill, S.C., Barber, P.W.: Scattered and internal intensity of a sphere illuminated with a Gaussian beam. IEEE Trans. Antennas Propag. **41**(3), 295–303 (1993)
25. Doicu, A., Wriedt, T.: Plane wave spectrum of electromagnetic beams. Opt. Commun. **136**, 114–124 (1996)
26. Barton, J.P., Alexander, D.R.: Fifth-order corrected electromagnetic field components for a fundamental Gaussian beam. J. Appl. Phys. **66**(7), 2800–2802 (1989)
27. Salamin, Y.I.: Fields of a Gaussian beam beyond the paraxial approximation. Appl. Phys. B **86**, 319–326 (2007)
28. Nieminen, T.A., Rubinsztein-Dunlop, H., Heckenberg, N.R.: Multipole expansion of strongly focussed laser beams. J. Quant. Spectrosc. Radiat. Transfer **79**(80), 1005–1017 (2003)
29. Richards, B., Wolf, E.: Electromagnetic diffraction in optical systems ii. structure of the image field in an aplanatic system. Proc. R. Soc. London, Ser. A **253**, 358–379 (1959)
30. Maia Neto, P.A., Nussenzveig, H.M.: Theory of optical tweezers. Europhys. Lett. **50**(5), 702–708 (2000)
31. Mazolli, A., Maia Neto, P.A., Nussenzveig, H.M.: Theory of trapping forces in optical tweezers. Proc. R. Soc. London, Ser. A **459**, 3021–3041 (2003)
32. Taylor, J.M., Love, G.D.: Multipole expansion of Bessel and Gaussian beams for Mie scattering calculations. J. Opt. Soc. Am. A **26**(2), 278–282 (2009)
33. Taylor, J.M., Love, G.D.: Optical binding mechanisms: A conceptual model for Gaussian beam traps. Opt. Express **17**(17), 15381–15389 (2009)
34. Chan, J.C.T., Sheng, P., Lin, Z.: Strong optical force induced by morphology-dependent resonances. Opt. Lett. **30**(15), 1956–1958 (2005)
35. Saad, Y., Schultz, M.H.: Gmres: a generalized minimal residual algorithm for solving nonsymmetric linear systems. SIAM J. Sci. Stat. Comput. **7**, 856–869 (1986)
36. Brevik, I.: Experiments in phenomenological electrodynamics and the electromagnetic energy-momentum tensor. Phys. Rep. **52**(3), 133–201 (1979)
37. Dienerowitz, M., Mazilu, M., Dholakia, K.: Optical manipulation of nanoparticles: a review. J. Nanophotonics **2**, 021875 (2008)
38. McGloin, D., Carruthers, A.E., Dholakia, K., Wright, E.M.: Optically bound microscopic particles in one dimension. Phys. Rev. E **69**, 021403 (2004)
39. Brevik, I., Sivertsen, T.A.: Radiation forces on an absorbing micrometer-sizes sphere in an evanescent field. J. Opt. Soc. Am. B **20**(8), 1739–1749 (2003)
40. Iver Brevik.: Private communication. (2008)
41. Pfeifer, R.N.C., Nieminen, T.A., Heckenberg, N.R., Rubinsztein-Dunlop, H.: Colloquium: Momentum of an electromagnetic wave in dielectric media. Rev. Mod. Phys. **79**(4), 1197–1216 (2007)
42. Woods, D., Mellor, C.D., Bain, C.D., Lewis, A., Ward, A.D.: Optical sculpting of emulsion droplets Progress in Electromagnetic Research Symposium. Hangzhou, China (2008)
43. Woods, D., Mellor, C.D., Taylor, J.M., Bain, C.D., Berry, M.G., Ward, A.D.: Optical nanofluidics. in preparation
44. Press, W.H., Teukolsky, S.A., Vetterling, W.T., Flannery, B.P.: Numerical Recipes in Cm. Cambridge University Press, Cambridge (1988)
45. Kawano, M., Blakely, J.T., Gordon, R., Sinton, D.: Theory of dielectric micro-sphere dynamics in a dual-beam optical trap. Opt. Express **16**, 9306–9317 (2008)
46. Meiners, J.C., Quake, S.R.: Direct measurement of hydrodynamic cross correlations between two particles in an external potential. Phys. Rev. Lett. **82**(10), 2211–2214 (1999)

47. Chen, J.C., Kim, A.S.: Brownian dynamics, molecular dynamics, and monte carlo modeling of colloidal systems. Adv. Colloid Interface Sci. **112**, 159–173 (2004)
48. Doi, M., Edwards, S.F.: The Theory of Polymer Dynamics International Series of Monographs on Physics. Clarendon Press, Oxford (1986)
49. Mellor, C.D., Bain, C.D.: Array formation in evanescent waves. Chem. Phys. Chem. **7**(2), 329–332 (2006)
50. Mellor, C.D., Fennerty, T.A., Bain, C.D.: Polarization effects in optically bound particle arrays. Opt. Express **14**, 10079–10088 (2006)
51. Xu, Y.-.L., Gustafson, B.A.S.: A generalized multiparticle mie-solution: further experimental verification. J. Quant. Spectrosc. Radiat. Transfer **70**, 395–419 (2001)
52. Y.-L. Xu. ftp://astro.ufl.edu/pub/xu/gmm01ff.
53. Apple Computer Inc. http://images.apple.com..
54. Jack, N.g., Lin, Z.F., Chan, C.T., Ping, Sheng.: Photonic clusters formed by dielectric microspheres: Numerical simulations. Phys. Rev. B **72**, 085130 (2005)
55. Šiler, M., Čižmár, T., Šerý, M., Zemánek, P.: Optical forces generated by evanescent standing waves and their usage for sub-micron particle delivery. Appl. Phys. B **84**, 157–165 (2006)
56. Almaas, E., Brevik, I.: Radiation forces on a micrometer-sized sphere in an evanescent field. J. Opt. Soc. Am. B **12**(12), 2429–2438 (1995)
57. Taflove, A., Hagness, S.C.: Computational Electrodynamics: The Finite-Difference Time-Domain Method. Artech House, Boston (2000)
58. Karásek, V., Brzobohatý, O., Zemánek, P.: Longitudinal optical binding of several spherical particles studied by the coupled dipole method. J. Opt. A **11**, 034009 (2009)

Chapter 3
Evanescent Wave Trapping

"...the totality is not, as it were, a mere heap, but the whole is something besides the parts..." Aristotle, Metaphysics, Book 8.

3.1 Introduction

This chapter considers the behaviour of microparticles in an evanescent wave trap. The effects of evanescent waves on microparticles was first investigated by Kawata and Sugiura in [1]. They found that particles were held on or close to the totally-internally-reflecting surface, and were transported along the surface by the transfer of momentum due to their scattered light.[1] This situation, with a single beam and single particle, attracted some theoretical interest, particularly in terms of the question of the interaction between the sphere and the surface [2–4]—a difficult question which has still not entirely been resolved.

Subsequently, interest has been extended to the trapping of particles in a counter-propagating evanescent field, where the momentum of the opposing beams cancels out, allowing stable trapping in two dimensions at the interface. An interesting effect was noted [5–7] whereby a single particle may be trapped on bright interference fringes or at the nodes between interference fringes, depending on the particle radius. This general effect in fact applies equally to any beam, and is not restricted to evanescent waves.[2]

Applications of evanescent wave trapping include particle transport and sorting [9, 10]. As is noted in those papers, optical binding interactions between nearby particles can have significant effects on the observed effects. This can be

---

[1] In the case of ideal total internal reflection from a planar interface, there is no energy or momentum transfer across the interface. However the presence of a particle on the far side of the interface alters the boundary conditions, leading to frustrated total internal reflection and hence momentum transfer to the particle. See Sect. 3.2 for further discussion

[2] Lekner considered the case of counter-propagating plane waves [8], although there do appear to be errors in the quantitative values quoted in that paper.

J. M. Taylor, *Optical Binding Phenomena: Observations and Mechanisms*,
Springer Theses, DOI: 10.1007/978-3-642-21195-9_3,
© Springer-Verlag Berlin Heidelberg 2011

Fig. 3.1 Examples of
"crystal" structures of
optically bound particles
from [12] (Mellor et al.)
(**a**) shows 520 nm
polystyrene particles in two
parallel-polarized beams, and
(**b**) shows the same particles
after the polarization of one
beam is rotated through 90°,
(**c**) shows a close-up of a
"broken hex" structure where
alternate fringes are
occupied. (All images from
Mellor et al. [12])

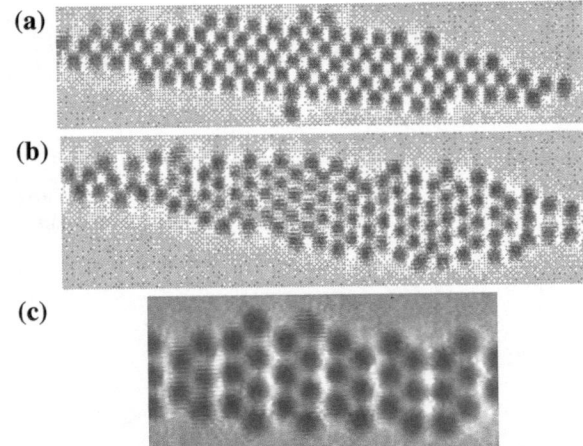

particularly prominent due to the broad size of the trapping/transporting beam, which means that particles some distance apart are illuminated with mutually coherent light, giving rise to the interference effects familiar in optical binding. This is in contrast to "optical tweezers" experiments where each particle is generally trapped using a separate laser, or using "time sharing" techniques. Here there will not be interference between the light scattered by each particle, and any optical binding effects will be very weak.

A number of groups have taken an interest in the optical binding of small microparticles in an evanescent wave trap, notably Bain, Mellor et al. [11, 12] at Durham, Zemanek et al. [10, 13] in the Czech Republic, and Ritchie et al. at Oxford. Impressive experimental results have been reported involving two-dimensional "crystal" structures. Examples of these are shown in Fig. 3.1. The effects observed included a range of different regular sphere packing structures, based on a range of hexagonal and "chessboard" lattices, and more complex structures with inhomogeneous spacings (as shown in Fig. 3.1c). Full details of the experimental setup and the investigations carried out can be found in [11, 12].

The theoretical simulations and graphs that appear in this chapter were all calculated using our Mie scattering model described in Chap. 2.

This chapter contains numerical simulation and interpretation of these experiments. We give theoretical explanations for the formation of chains and two-dimensional clusters in an evanescent wave trap, and make some experimentally-verifiable predictions of unexpected aspects to this behaviour. We find that in many cases the formation of regular "crystal" structures is due to the interplay between optical and collisional interactions between the trapped particles. The experimental results reported in this chapter were obtained by a number of different groups (attributions are given in the figure captions). The numerical results and theoretical interpretations are all our own. Some of the content of this chapter has been published in [5].

We will briefly describe the experimental setup in the next section, before discussing the behaviour of single particles in a coherent counter-propagating evanescent wave trap (Sect. 3.3), and then extending this to one- and two-dimensional clusters (Sects. 3.5 and 3.6), and comparing the simulation results with the experimental results. The wide range of potential further investigation is outlined in Sect. 3.7. Various novel applications of GLMT to evanescent wave traps are contained in the Appendices to this chapter.

3.2 Evanescent Wave Trap

As described in Sect. 2.4.3, when a plane (or near-plane) wave encounters a planar dielectric boundary, approaching from the medium with higher refractive index, it may be totally internally reflected at the boundary. All the incident power is reflected at the interface, and a short-range evanescent field develops on the far side of the interface. In this ideal case, there is no energy flow across the boundary. However, if a particle is present on the other side of the boundary then this disturbs the symmetry of the interface. The particle can scatter light into the medium containing the evanescent field, in an effect similar to frustrated total internal reflection [14, p1200]. This scattering effect means that the field can exert forces on the particle just as it would in a non-evanescent situation.

A particle in the evanescent field of a single beam will be pushed along the boundary in a direction parallel to the propagation direction of the incident field [1]. In order to achieve stable trapping, two counter-propagating beams are required [11, 12] so that the net lateral force on a particle is zero. The experimental setup used is shown in Fig. 3.2. A single laser beam is reflected at the surface of a prism (beyond the critical angle), and this beam is then retro-reflected so as to provide the counter-propagating beam. Polystyrene particles between 200 nm and 800 nm in diameter are introduced into the water above the prism surface, and interact with the light field and with each other. The particles are imaged from above using a microscope objective. The experimental parameters, which were used in the simulations discussed throughout this chapter, are listed in Table 3.1.

There are two main electromagnetic forces acting on the particles. Firstly, each individual particle will interact with the background evanescent light field. The nature of this interaction will depend on the relative polarizations of the two evanescent waves. If the two beams can be considered as incoherent then the only effect will be a weak gradient force which draws a particle into the centre of the beam, where the evanescent field intensity is highest. This is also approximately the case where the beams are orthogonally polarized, in which case there is little interference between the two beams. Conversely if the beams have the same polarization then there will be a standing wave generated by the interference of the two beams. This will form fringes with periods of the order of half the wavelength of the laser light. In this situation the dominant effect on an isolated particle is its

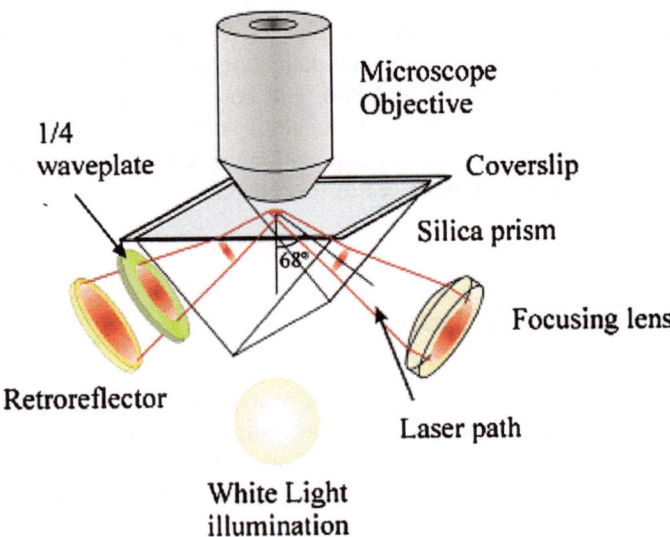

Fig. 3.2 Diagram (from [12]) of the experimental setup of a counter-propagating evanescent wave trap. A loosely-focused Gaussian beam is totally internally reflected at a silica-water interface on the surface of a prism. The reflected beam is retro-reflected by a mirror to provide the counter-propagating beam. This results in a region of order 50 μm in diameter where there is an evanescent wave structure generated in the water at the surface of the prism. Within this region, particles are trapped and can interact with each other through the electromagnetic field

	Parameter	Value	Comment
Table 3.1 Parameters used in the calculations in this chapter. In addition we refer to the particle radius a and the size parameter ka where k is the laser wavenumber in water	Vacuum wavelength	1064 nm	Nd:YAG laser
	Laser beam power	300 mW	
	Focal spot radius	8 μm	
	Substrate refractive index	1.45	Silica prism
	External refractive index	1.32	Particles in water
	Critical angle	65.6°	
	Angle of incidence	67°	
	Particle refractive index	1.57	Polystyrene

interaction with these fringes. This interaction contains a number of interesting subtleties, and is discussed in the next section.

The second interaction is that between multiple nearby particles in the evanescent trap. As in any optical trap, the particles will interact with each others' mutually scattered light, in a way which may differ considerably from the behaviour of isolated particles. This interaction is discussed later in this chapter.

3.3 Single Particles in Evanescent Fields

The behaviour of a single particle in the presence of interference fringes has previously been considered by Ng and Chan [7] in a GLMT calculation like our own, as well as by Siler et al. [6]. It is interesting to consider the behaviour as a function of particle size. Because of the strong modulation of the evanescent field intensity on scales smaller than a wavelength (in the case of parallel-polarized counter-propagating fields), the particle will interact strongly with the field through the gradient force.[3] Figure 3.3 shows the interaction of a single particle with the interference fringes. It can be seen that a Rayleigh particle (a particle much smaller than the wavelength of light) will be attracted to a region of highest intensity—in other words to a fringe maximum.

For larger particles, the situation is a little more complex. The Born approximation assumes that a particle does not significantly modify the field by its presence, which is a reasonable assumption in the case of a particle whose refractive index is close to that of the surrounding medium, even in the case of relatively large particles. Under the Born approximation, the behaviour of the particle can be determined by considering the background intensity integrated over the volume of the particle. Alternatively, a full Mie scattering calculation can be performed for the particle (whether or not the Born approximation is satisfied) but, as discussed in Appendix A, the Born approximation can still be useful in analysing the behaviour.

Figure 3.3 shows the results of such a Mie scattering simulation and shows the force on a single particle in a set of interference fringes as a function of particle size parameter ka. When the electric field of the laser is in the trapping plane, this is designated "S" polarization. Similarly "P" polarization is when the polarization is parallel to the plane of incidence. The plot shows that a single, small particle is attracted to bright fringes. At larger radii the particle's centre can instead be attracted to a dark region between fringes.

For a particle whose refractive index is close to that of the surrounding medium, this can also be understood through a gradient potential argument in the Born approximation. For some sizes of particle, the integrated intensity within the volume of the sphere is maximized when the particle is centred on the dark fringe *between* two bright fringes [6], because two fringes are fairly well covered by the particle, instead of one fringe being very well covered (see Fig. 3.3). We designate the radius at which the behaviour first switches from light-seeking to dark-seeking as the "crossover radius".

The distribution of the background electromagnetic field will of course vary as a function of the angle of incidence of the beam. As this angle is increased beyond the critical angle, the evanescent decay length of the field will decrease, and the period of the fringes will also decrease. One might then expect that the combination of these two effects would alter the crossover radius for a particle.

---

[3] The gradient potential and associated force are discussed in Sect. 2.6.2.

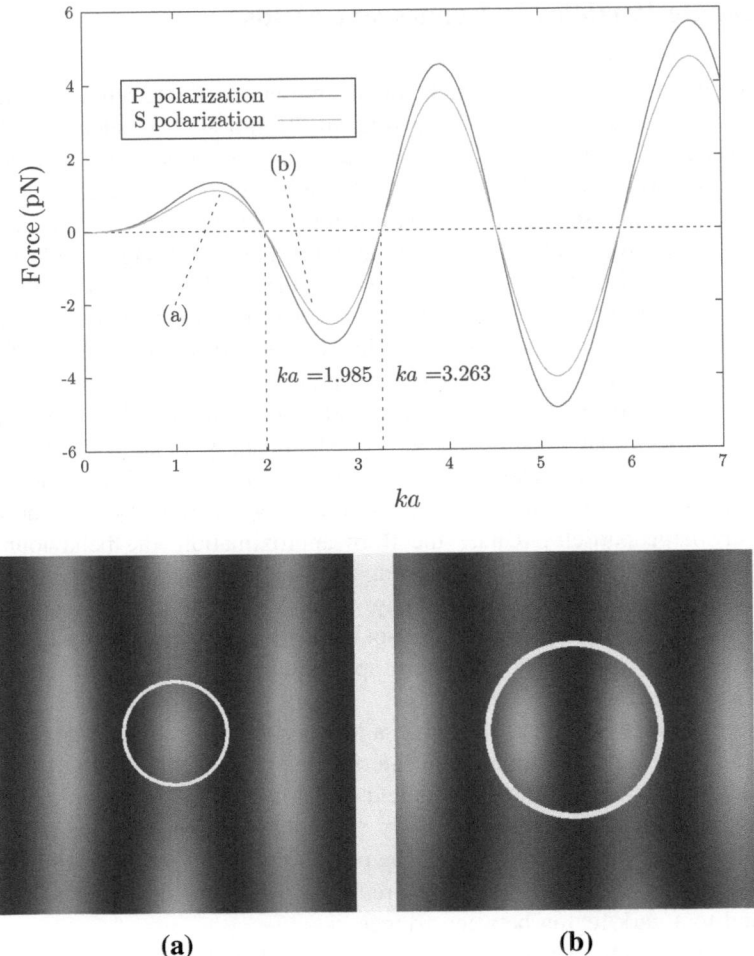

Fig. 3.3 The top graph shows the force acting on a single particle placed halfway between a bright and dark fringe, as a function of size parameter ka. A positive force indicates that the particle is attracted to the *bright fringe*. Two lines are shown for different polarization states. The first crossover occurs at $ka = 1.985$. For full parameters used, see Table 3.1. It can be seen that for a range of values of ka the force is negative and and therefore attracted to darker regions. The "crossover" sizes (where the particles switch from being attracted to light to dark) are indicated numerically on the graph. The *lower* two plots show simulated images showing how the interference fringes are distorted by the particle and that the particle sits on a *bright fringe* when small (**a**), and on a *dark fringe* when larger (**b**), The *white circles* indicate the particle size and location, and the sizes for (**a**) and (**b**) are indicated on the graph

Intriguingly, it turns out that these two effects exactly cancel each other out, such that the crossover radius is exactly the same regardless of the angle of incidence in the substrate. This completely unexpected result is discussed further in Appendix A.

3.4 Long-Range Nature of Optical Binding

At this point it is worth discussing a feature of optical binding which is critical to the effects seen in this chapter. Optical binding is a long-range interaction, where the *force* between two scatterers decreases in inverse proportion to the distance between them $(f \sim 1/r)$. This is in contrast to most physical effects (for example an electrostatic or gravitational interaction), which decrease following an inverse square law. This fact was highlighted and derived in [15], but theirs was a mathematical derivation which does not give much physical insight into the reasons behind the long-range nature of the interaction. Here we give a slightly less rigourous explanation, but one which reveals the physical origin of the long-range force.

Consider the scattered wave from a single particle. The amplitude of this wave decays as $1/r$ and hence its intensity decays as $1/r^2$. The gradient force on a second particle lying outside the illuminating beam is proportional to the gradient of intensity, and will therefore be proportional to $1/d^3$, where d is the distance between the two particles.

If, on the other hand, a second particle is also inside the illuminating beam, the scattered light interferes with the background laser light, forming fringes. A simple example of this is shown in Fig. 3.4. The fringe amplitude varies as $1 \pm \alpha/r$ (where α is a measure of the level of scattering by the particle). Thus their intensity varies as

$$(1 \pm \alpha/r)^2 \sim 1 \pm 2\alpha/r, \tag{3.1}$$

where we have for now made the approximation that the scattered field is a small perturbation to the laser field. Since the gradient force is proportional to intensity, we find that in addition to the effect of the external beam there is a force acting on the second particle whose strength varies as $1/d$. Although the figure compares examples of tightly-focused optical tweezers and a more loosely focused Gaussian beam, this simple argument applies to any broad, coherent beam illuminating multiple particles simultaneously.

A consequence of this is that, even for a one-dimensional chain, the magnitude of the forces on the central particle can grow indefinitely (logarithmically) with the length of the chain. To show this we will now extend our argument to consider a chain of $2n + 1$ particles, with an inter-particle spacing d chosen such that for a particle i at position id the scattered waves from every other particle are in phase. For large n the strength of the forces on the central particle will scale as

$$|F| \sim 1 + \sum_{i=1}^{n} \frac{2\alpha}{id} \sim 1 + \frac{2\alpha(\ln n + \gamma)}{d}, \tag{3.2}$$

where $\gamma = 0.5772...$ is the Euler-Mascheroni constant [16, (6.1.3)]. The force is a logarithmically-increasing function of n.

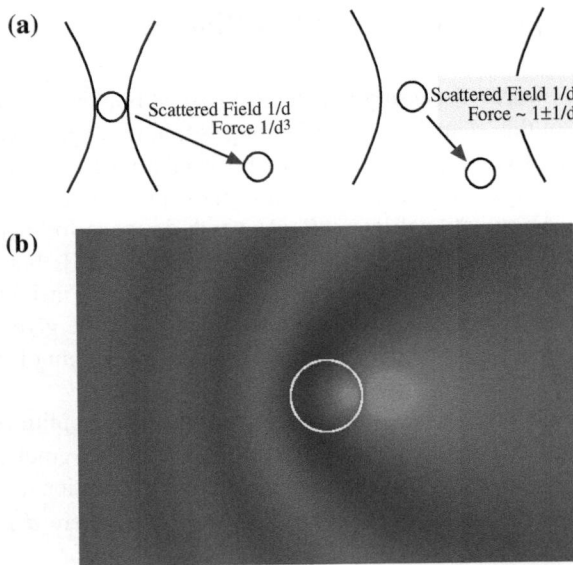

Fig. 3.4 **a** Schematic showing how the scattered intensity of light from a single particle (*left*) in a beam depends on $1/r$, and therefore the gradient force on the second particle outside the beam depends on $1/d^3$ for particle separation d. When both particles are exposed to the laser field (*right*), the modulation in the force scales with $1/d$. **b** Simulation showing the field intensity distribution caused by interference of a scattered wave from a particle with the evanescent field of a single laser beam propagating *left to right*, showing the production of interference fringes which lead to optical binding. (**a**) Optical binding scaling (**b**) Interference of scattered wave and laser field

This is in contrast to an electrostatic interaction between particles of charge q (and its associated inverse square law) which would asymptotically approach a constant value at large n. For large n the strength of the forces on the central particle will converge to a constant value:

$$|F| = \frac{1}{4\pi\epsilon_0} \sum_{i=1}^{n} \frac{q^2}{(id)^2} \sim \frac{1}{4\pi\epsilon_0} \frac{\pi^2 q^2}{6d^2}. \qquad (3.3)$$

Thus (as might intuitively be expected) in this case the effects of distant particles are negligible and the force converges to a constant value at large n.

Returning to the gradient force, for a two-dimensional cluster the strength of the forces will grow in proportion to the radius of the cluster. This is an important result which is central to many-particle optical binding, and has a number of consequences:

- The asymptotic behaviour of a large cluster of spheres cannot be easily predicted by extrapolating the behaviour of smaller clusters of spheres.

- The shape of the surface of a cluster of particles can have a strong effect on the forces on particles at the core of the cluster.
- As the number of particles in a cluster grows, the influence of the background field will become proportionally weaker until the structure of the cluster is almost entirely determined by the inter-particle interactions, irrespective of the background field distribution.

3.5 Anomalous Behaviour of Chains

Next we will consider the force on a particle within a chain of particles parallel to the interference fringes, as illustrated in Fig. 3.5. In this case we find that the behaviour can be completely different from that of individual particles. Various particle radii were investigated, exposed to both S and P polarization states. The results are summarized in Fig. 3.6, which shows the force acting on the central particle in a chain halfway between a fringe maximum and a fringe minimum (in analogy to Fig. 3.3). From this we determine whether the chain is attracted to a bright or dark fringe and the strength of that attraction.

It can be seen that in P-polarized light a large chain of particles is attracted to bright fringes despite the fact that a single particle is attracted to dark fringes. The force acting on an individual particle in the chain grows with the size of the chain. Figure 3.7 shows the scattering behaviour in various situations, which leads to field distributions such as that shown in Fig. 3.5 when P-polarized light is used, and explains the different behaviour for S and P-polarized light.

There is competition between the external optical landscape of the fringes, which in this case attracts an individual particle towards dark regions, and the scattering behaviour of the ensemble of particles which tends to attract the ensemble towards bright fringes. The force due to the interference fringes is smallest close to the "crossover radius", and hence a chain of only 3 particles of radius 260 nm is able to overcome the influence of the "landscape" and settle on a bright fringe. For 280 nm particles, a chain of 13 particles is required to overcome the increased force of the background landscape.

Figure 3.8 compares our numerical results with the prediction of (3.1). This supports our description of optical binding as a long-range interaction scaling with $1/r$, but shows that the picture is complicated by multiple scattering: optical binding has become the dominant influence on the optical landscape, and the perturbation approach taken earlier is no longer entirely appropriate. Nevertheless, the first-order approach can offer some strong insights into the complex nonlinear behaviour of the system.

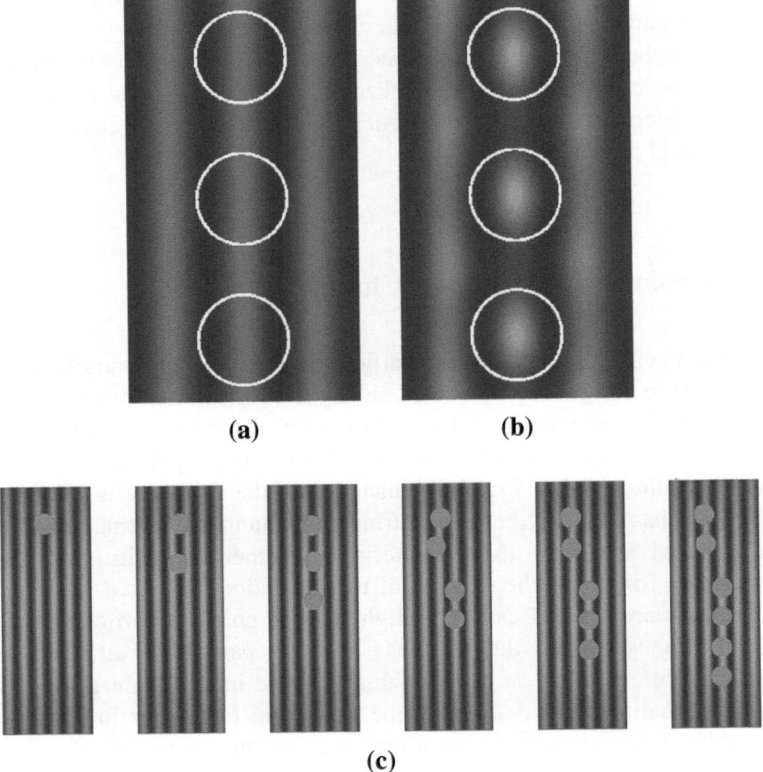

Fig. 3.5 Examples of chains of particles along the fringe direction. (**a**) is shown without scattered light, for clarity (**b**) shows how the fringe pattern can be modified by the presence of the spheres (**c**) Frames from a Brownian motion simulation which demonstrates the switch in fringe affinity for 265 nm radius particles as more particles are manually added to the system. The particles are free to move in three dimensions in this simulation. When the fourth particle is added to the simulation, the particles switch from being attracted to *dark fringes* to being attracted to *light fringes*. In this case the random Brownian motion of the particles has led to two particles going *left* and two particles going *right*

3.6 2D Clusters and Comparison with Experiment

We are able to reproduce the "broken hex" structure shown in (Fig. 3.1c) in numerical experiments, along with the initially surprising conclusion that the particles are located close to fringe maxima, despite being of a radius where a single particle would be attracted to a fringe minimum.[4] Figure 3.9 shows

---

[4] This finding was the inspiration for the work on one-dimensional chains of particles discussed in Sect. 3.5.

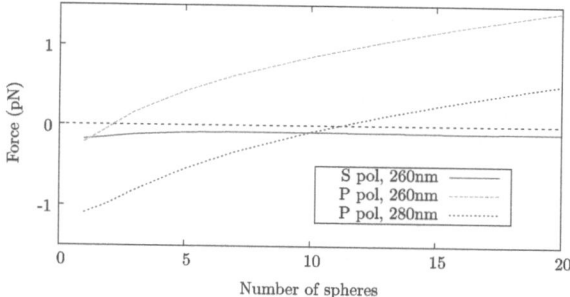

Fig. 3.6 Lateral force acting on the central particle of a chain of particles as a function of number of spheres in the chain. The chain is in longitudinal equilibrium, but has been constrained to a lateral position halfway between a *light* and a *dark* fringe in order to show its light- or dark-seeking behaviour. The forces are shown for both S and P laser polarizations for spheres of radius 260 nm, and for P polarization for spheres of radius 280 nm. A positive force indicates that the particle is attracted to the *bright fringe*, and a negative force a *dark fringe*. For full parameters used, see Table 3.1

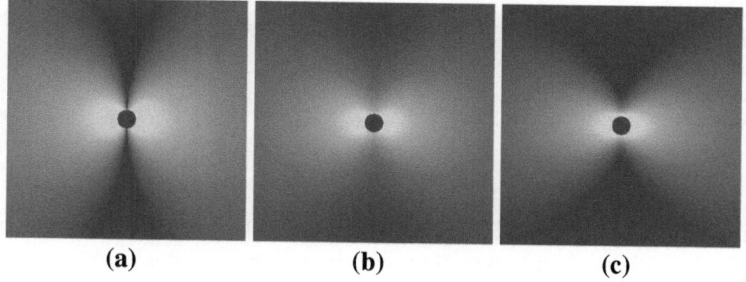

Fig. 3.7 Scattered field intensity in the trapping plane for a particle with $ka = 2.5$. The intensity is displayed on a logarithmic scale. **a** P-polarized beams and particle on a *dark fringe*. There is no scattering in the direction of the fringes when a particle lies on a dark fringe. **b** P-polarized beams and particle on a light fringe. A particle situated on a bright fringe formed by P-polarized light scatters relatively strongly in the direction of the fringes. This scattering enhances the intensity of the fringe that the particle lies on, which makes it more energetically favourable for other particles to be situated on that same fringe. **c** S-polarized beams and particle on a light fringe. There is very weak far-field scattering in the direction of the fringes (a dipole scatterer will not scatter in a direction parallel to the electric field polarization)

a Brownian motion simulation of such a structure. Investigation using Brownian motion simulations has revealed the origin of these structures:

- A chain of particles orientated parallel with an interference fringe will position itself on a fringe maximum, for the reasons discussed in Sect. 3.5.
- The particles are of a physical size such that they cannot occupy every adjacent interference fringe.
- In the case of counter-propagating evanescent plane waves, there is only very weak interaction between particles on nearby fringes, and the chains on

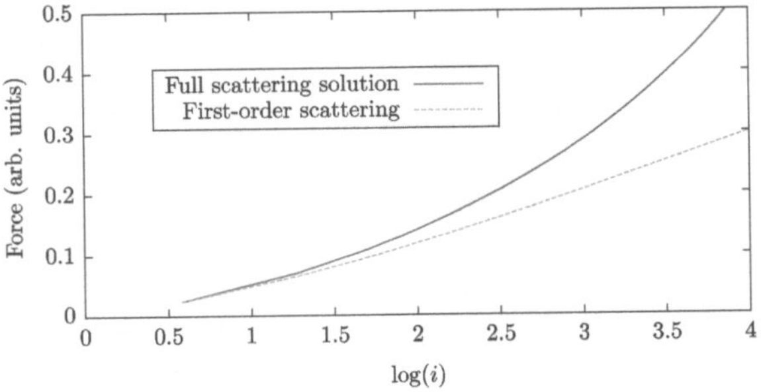

Fig. 3.8 Lateral force acting
on the central of $(2i + 1)$
particles. Equation 3.2
predicted a logarithmic
relationship, but this is not
seen for the full solution. If
we consider just the first-
order scattered field though
(see Sect. 2.5.1), the relationship is clear

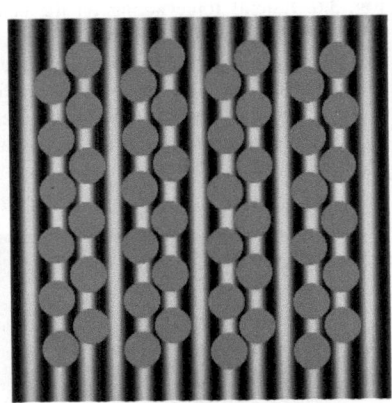

Fig. 3.9 Screenshot from a
full Brownian motion
simulation of a "broken hex"
structure of trapped spheres.
The particles are located
close to fringe maxima,
despite being of a radius
where a single particle would
be attracted to a fringe
minimum. The Brownian
motion simulation confirmed
the meta-stability of this
cluster structure

individual fringes tend to wander apart through random Brownian motion,
physical collisions and any electrostatic repulsion which may be present.
However, if a broad trapping potential is introduced, caused by the broad
intensity variation of a loosely-focused Gaussian beam reflected at the interface, then there is a force which pushes individual chains towards the centre of
the trap. The net result is that the lowest-energy configuration sees one fringe in
three left vacant, as shown in the experimental and numerical video frames.

This investigation has revealed that in this close-packed environment it is
important to consider interactions other than the light-mediated inter-particle
forces—in particular, Brownian motion and electrostatic repulsion. Unfortunately
this presents a significant challenge for modeling of the system, since the modeling
of Brownian motion demands very small timesteps to be used in the simulation,
and physical contact between spheres is difficult to model and makes calculation of
the electromagnetic field more difficult.

A further illustration of the importance of interactions other than optical
binding is given by an experiment performed by Bain et al. [12] where the
polarization of both beams was changed from P to S. This alters the vector
components of the evanescent field, and leads to a change in cluster structure.
Although the effect is far better illustrated in a video, Fig. 3.10 shows the shape of
the cluster before and after the switch. With P polarization the particles are not in

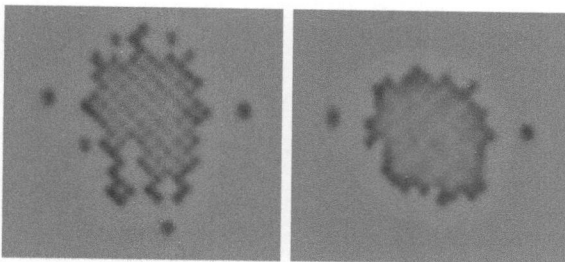

Fig. 3.10 Particle cluster formed with interfering P-polarized evanescent waves (*left*) and after polarization is changed to S (*right*). The cluster collapses in the *vertical* direction until the particles are in contact with each other. This is a spacing determined by physical contact between spheres, rather than by light-mediated forces. Consequently, there is *not* strong optical binding in this configuration, and the particles on the edges of the cluster are significantly more mobile under Brownian motion. (Bain et al. private communication)

contact, and their equilibrium spacing is determined by light-mediated forces. Because they assume an inter-particle spacing which minimizes their energy (gradient potential), the result is a strong enhancement of the light intensity within and around the particles, making this a relatively strongly-bound configuration. We observed that the video shows that the particles are bound strongly enough that there is very little Brownian motion of the individual particles, although there is some bulk motion of the entire cluster. In contrast, with S polarization the particles are in contact with each other. The inter-particle spacing is determined by physical contact, rather than by the light-mediated forces. Thus there is no particular enhancement of the intensity in and around the particles through constructive interference, and the light forces are relatively weak. Consequently the video shows considerable Brownian motion of individual particles around the perimeter of the cluster, since the constituents of the cluster are bound together much less strongly.

3.7 Conclusions and Further Work

In this chapter we have considered optical binding of clusters in an evanescent wave trap. We have found that:

- The behaviour of a chain of particles can be completely different from the behaviour of an isolated particle.
- Two-dimensional structures such as the "broken hex" structure are reproduced in numerical simulations, which make unexpected predictions about the location of the particles relative to the interference fringes—an effect which could be tested in future experiments.
- The nature of the forces on an isolated particle does not depend on the angle of incidence in the substrate material.

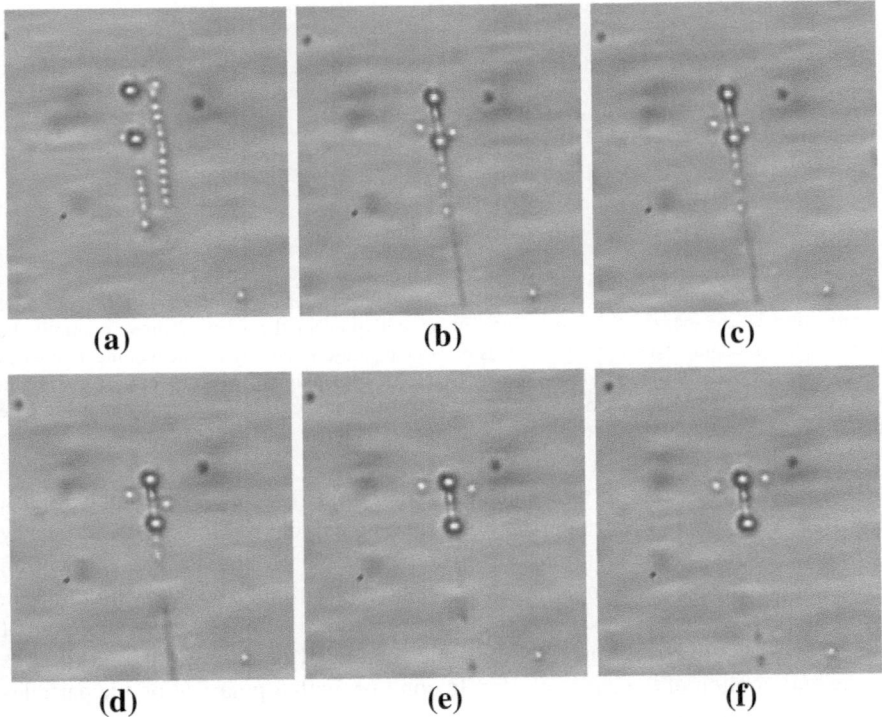

Fig. 3.11 Experimental video of "photonic oxygen" formation (G. Ritchie et al. University of Oxford, private communication). Frames at approximately 250 ms intervals. Particles of two different diameters are mixed in the trap. The smaller particles form chains of touching particles (*parallel* with the direction of propagation of the trapping beams; frame **a**), but it seems to be energetically favourable for the chains to be capped by larger particles (frames **e-f**). If this occurs, any additional small particles on the end of the chains are violently repelled from the vicinity of the chain (frames **b-d**)

- In general, the forms of particle clusters in an evanescent trap are determined by a combination of optical binding effects and physical collisions between spheres.

There is considerably more experimental and theoretical work which could be carried out on evanescent wave trapping. Examples which deserve further attention include:

- Further investigation of the dependence of the particle clustering on the properties of the particles (such as refractive index and diameter) and on the properties of the beams (such as their polarization). It would be interesting to reach more general conclusions on the relative importance of optical binding and physical collisions between particles in determining the cluster shapes.
- Modeling of the experimental results obtained by Grant Ritchie's group in Oxford (see Fig. 3.11). In their experiment, particles of two different radii were mixed together, and they tended to form combined structures reminiscent of

diagrams of bi-atomic molecules such as oxygen, which they dubbed "photonic oxygen". Here it seems to be preferred for larger spheres to cap a chain of smaller spheres, with any further small spheres being strongly repelled from the ends of the capped chain.

- Experimental demonstration of whether particles are trapped on fringe maxima or fringe minima. This should in theory be possible by looking at the angular dependence of the light scattered by an individual particle and/or a cluster of particles. However preliminary investigations by Pavel Zemanek's group suggest that this is far from easy in a noisy experimental environment.
- Modeling of the trapping of gold nanoparticles. This is something that Colin Bain's group have investigated. They observed quasi-stable, but apparently random, configurations which were stable for a few seconds before there was very rapid motion of all the particles, leading to a new quasi-stable configuration. It would be interesting to investigate this with our computer model, and to establish whether this quasi-stable behaviour is caused by optical binding, or whether it may have other origins such as mode-hopping within the laser.
- Further investigation of the lack of angular dependency in evanescent wave trapping (see Appendix A).
- It would be very interesting to consider the effects of scattering from the substrate on the behaviour of trapped particles: this may well be very important in the case of multiple particles. This is discussed in Appendix B.

Appendix A: Beyond the Critical Angle

In the simulations discussed in this chapter, the angle of incidence has generally been close to the critical angle. It is natural to wonder how behaviour such as the light-or dark-seeking behaviour of a single particle (as shown in Fig. 3.3) changes as the angle of incidence of the beam (and hence the characteristics of the evanescent field) vary. As the angle of incidence increases, the fringe spacing decreases, which would be expected to decrease the values of the "crossover radii". However, the decay length of the evanescent field will also decrease. This means that for a sphere lying on top of the prism, the intensity is greater close to the bottom of the sphere, where the sphere is narrower. Qualitatively this should lead to an increase in the crossover radius since the "effective size" of the sphere is smaller in that area of highest intensity closest to the interface. Fascinatingly, we found via simulation that this crossover radius is completely unaffected by the angle of incidence. We therefore attempted to prove this hypothesis: in the Born approximation the crossover radius for a particle is independent of the angle of incidence. This hypothesis has interesting implications for evanescent wave trapping, Although we have not achieved a full proof of the result, there is overwhelming evidence that it holds. In this Appendix we will outline our attempted route to the proof.

The electric field of the evanescent wave has the following form:

$$E = e^{-\beta z} e^{iukx} \tag{3.4}$$

where $u = \cos(\theta_k) = \frac{n_s}{n_{ext}} \sin(\theta_i)$ defines the complex analogue of the angle of transmission for an evanescent wave produced by a wave incident at angle θ_i, and $\beta = k\sqrt{u^2 - 1}$. In the Born approximation we can calculate the gradient potential of a particle of radius a in counter-propagating S-polarized evanescent fields using the following integral (see (2.36)):

$$U_{grad} = -\frac{\epsilon_0}{4} (n_{int} - n_{ext}) \int\limits_V 2|E_{cp}|^2 \, dV \tag{3.5}$$

$$E_{cp} = \frac{1}{2} e^{-\beta z} (e^{iukx} + e^{-iukx})$$

The integral is evaluated over the volume of the sphere, whose centre lies at $z = a$ such that the sphere is sitting on the interface.

In order to simplify the analysis, we will take out a leading factor of $e^{-2\beta a}$. This acts as a scaling factor which modifies the potential as a function of the sphere radius. It is easier, however, to take this factor out and perform the integral around the origin; and as we will see later, with this factor removed the remaining integral (which represents a sphere whose centre is at a fixed z coordinate regardless of its radius a) is *independent of* θ_i. If we do this, evaluate $|E_{cp}|^2$, and write x and z in polar coordinates, we have a new potential without this leading factor:

$$U'_{grad} = \int\limits_0^{2\pi} \int\limits_0^{\pi} \int\limits_0^{a} e^{-2\sqrt{u^2-1}kr\cos\theta} \cos(kru \sin\theta \cos\phi) r^2 \sin\theta \, dr \, d\theta \, d\phi \tag{3.6}$$

If we can show that this potential U'_{grad} is independent of u for the crossover radii, then we will have shown that the crossover radius is independent of the angle of incidence. In fact, we will attempt to prove the more general result that U'_{grad} is independent of u for *any* radius. This is equivalent to saying that, aside from the leading exponential factor of $e^{-2\beta a}$, the interaction of *any* sphere with the evanescent field is independent of the angle of incidence. The advantage of tackling this more general result is that we can eliminate the radial integral: as long as we know the condition holds in the limit of small radius, $a \to 0$, then all we need to prove is that $\frac{\partial U'_{grad}}{\partial a}$ is independent of u, in other words that the following 2D integral:

$$I = \int\limits_0^{2\pi} \int\limits_0^{\pi} e^{-2\sqrt{u^2-1}kr\cos\theta} \cos(kru \sin\theta \cos\phi) \sin\theta \, d\theta \, d\phi \tag{3.7}$$

is independent of u for all r. We can solve the integral with respect to ϕ :

$$I = 2\pi \int_0^{\pi} e^{-2\sqrt{u^2-1}kr\cos\theta} J_0(2kr\cos(kru\sin\theta))\sin\theta\,\mathrm{d}\,\theta. \qquad (3.8)$$

It does not appear to be possible to solve this integral analytically; certainly the software package Mathematica is unable to do so. It may still be possible, however, to show that the result is independent of u for all r, by expanding the exponential and Bessel function as a Taylor series, and treating individual powers of u and θ. As yet, we have not been successful with this, but it is possible to show to many decimal places that for a range of sample values of u and r the result of this single integral is independent of u. There is thus overwhelmingly strong evidence that this intriguing result is true, even if we have not yet been able to prove this analytically.

Appendix B: Force on a Sphere on a Substrate

In all the models discussed in this chapter, we have assumed that the effects of the substrate/water interface can be ignored. We have treated the particles as confined to the plane $z = a$ (where a is the particle radius), but we have surrounded them by a uniform external medium (the water) without any reference to the interface. For the case of a single particle, this assumption was supported by ray-optics investigations carried out in [2]. Light reflected by the boundary *which is then incident again on the sphere* will be reflected at acute angles, where there is a relatively weak Fresnel reflection coefficient. Consequently, this light will have little effect on the field scattered *by the sphere*, or on the force on the sphere. There may however be significant effects on the far-field scattering pattern produced by the combined sphere/interface system.

In the case of multiple particles, it is less clear that this assumption holds. Light scattered by one sphere can be reflected by the interface and then be incident on a *different* sphere. In such cases, the angle of reflection at the boundary may be very oblique, in which case there will be a relatively strong Fresnel reflection from the boundary, and this reflected light cannot be ignored in a full treatment of the system. It is possible that this is one of the reasons that it has been difficult to demonstrate good agreement between simulation and experiment for some evanescent trapping experiments involving multiple particles.

Unfortunately, it is far from trivial to include the effects of the substrate in a Mie scattering calculation. Fundamentally, this is because GLMT is designed to be mathematically suited to spherical symmetry, which the interface does not possess. A number of authors have calculated the light scattering by a single particle on a substrate [17–19]. Their various approaches to this problem all involve the numerical calculation of a number of complicated integrals.[5]

---

[5] For a perfectly-reflecting boundary, a technique analogous to the method of *image charges* in electrostatics can be used. However for a dielectric boundary this approach cannot be used.

However, we again emphasize that calculating the force on a sphere on a substrate is *not the same* as calculating its far-field scattering pattern: to calculate the far-field scattering we must consider all light scattered from the sphere and then re-scattered from the surface. This means that the scattered light is coming from a region of sources which is effectively of infinite extent (all points on the plane of the interface). Conversely, in order to calculate the force on the sphere we only need to consider light scattered from the interface which is *then incident again* on the sphere. Thus the region of sources which we need to consider is relatively compact (of the order of twice the size of the sphere), and all we need for the force calculation is to know the actual external field *on the surface* of our sphere.

What we have is effectively a multiple scattering problem between the substrate and the sphere, analogous to the problem with two spheres. Indeed, with a little care we can consider the substrate as the limit of an infinitely large sphere, which allows the problem to be treated within the framework of GLMT. The simple problem of two touching spheres of different radii exposed to an external field $\mathbf{a}_{ext}$ is described by (2.23) (but for the two particle case). This is not quite what we need here, though. If we used that equation, we would include the non-physical effect of the background field being scattered by a large sphere which represents the substrate, whereas in fact the external (evanescent) field has already considered that interaction, which has given rise to the evanescent field in the first place. However, we *do* need to correctly consider the interaction of the scattered light from sphere (1) with the surface of the "substrate" sphere (2). These requirements can be met if (from an order-of-scattering viewpoint) we arrange that sphere (2) does not "see" the zero-order field, but only the higher-order scattered fields. This results in the following matrix equation:

$$
\begin{bmatrix} \mathbf{a}^{(1)} \\ \mathbf{a}^{(2)} \end{bmatrix} = \begin{bmatrix} \mathbf{a}_{ext}^{(1)} \\ 0 \end{bmatrix} + \begin{bmatrix} 0 & \mathbf{F}_{21} \\ \mathbf{F}_{12} & 0 \end{bmatrix} \begin{bmatrix} \mathbf{s}^{(1)} \\ \mathbf{s}^{(2)} \end{bmatrix} \tag{3.9}
$$

which differs from (2.23) only by the absence of the $\mathbf{a}_{ext}^{(2)}$ term. By solving this equation system we can arrive at a very simple result: a modification to the scattering matrix $\mathbf{T}$ for sphere (1). If this problem was solved for the full scattered field $\{\mathbf{s}^{(1)}, \mathbf{s}^{(2)}\}$ by a naive inversion (as in Sect. 2.5.2), we would require an inordinately large value of n_{max} to describe the field at sphere (2), but clever algebraic manipulation can avoid this requirement, exploiting the fact that we do not actually directly require $\mathbf{s}^{(2)}$, and describing each element of the new $\mathbf{T}$ matrix in terms of an infinite sum. Although these infinite sums must be truncated at a relatively large value of n_{max}, we only require this for the relatively small number of terms in the single-sphere $\mathbf{T}$ matrix, which keeps the calculation tractable.

There is no reason this approach cannot be extended to scattering between multiple spheres, where it would be much more useful (since this is where substrate effects are expected to be most significant, due to higher angles of incidence with the substrate). The aim would be to modify the translation matrix $\mathbf{F}_{ij}$ between two spheres i and j, expressing each element in it as an infinite sum in the same way as the modification to the single-sphere **T** matrix.[6] This would be computationally fairly difficult to calculate, and its performance would have to be compared to the integral approaches taken by other authors. It is very possible that this approach would be computationally faster, though, and it would certainly be easier to implement correctly in computer code, since it requires only a very slight modification to the existing GLMT approach. While such an approach is unlikely to be fast enough for use in a time-evolution simulation, it could be used to investigate individual cases and establish whether such effects cause significant modification to the binding behaviour or not.

References

1. Kawata, S., Sugiura, T.: Movement of micrometer-sized particles in the evanescent field of a laser beam. Opt. Lett. **17**(11), 772–774 (1992)
2. Prieve, D.C., Walz, J.Y.: Scattering of an evanescent surface wave by a microscopic dielectric sphere. Appl. Opt. **32**, 1629–1641 (1993)
3. Almaas, E., Brevik, I.: Radiation forces on a micrometer-sized sphere in an evanescent field. J. Opt. Soc. Am. B **12**(12), 2429–2438 (1995)
4. Brevik, I., Sivertsen, T.A.: Radiation forces on an absorbing micrometer-sizes sphere in an evanescent field. J. Opt. Soc. Am. B **20**(8), 1739–1749 (2003)
5. Taylor, J.M., Wong, L.Y., Bain, C.D., Love, G.D.: Emergent properties in optically bound matter. Opt. Express **16**, 6921–6929 (2008)
6. Šiler, M., Šerý, M., Čižmár, T.: Pavel Zemánek (2005) Submicron particle localization using evanescent field. In: Proceedings of the SPIE, **5930**, p. 59300R (2005)
7. Ng, J., Chan, C.T.: Private communication (2006)
8. Lekner, J.: Force on a scatterer in counter-propagating coherent beams. J. Opt. A **7**, 238–248 (2005)
9. Čižmár, T., Šiler, M., Šerý, M., Zemánek, P., Garcés-Chávez, V., Dholakia, K.: Optical sorting and detection of submicrometer objects in a motional standing wave. Phys. Rev. B **74**, 035105 (2006)
10. Šiler, M., Čižmár, T., Šerý, M., Zemánek, P.: Optical forces generated by evanescent standing waves and their usage for sub-micron particle delivery. Appl. Phys. B **84**, 157–165 (2006)
11. Mellor, C.D., Bain, C.D.: Array formation in evanescent waves. Chem. Phys. Chem. **7**(2), 329–332 (2006)
12. Mellor, C.D., Fennerty, T.A., Bain, C.D.: Polarization effects in optically bound particle arrays. Opt. Express **14**, 10079–10088 (2006)
13. Šiler, M., Zemánek, P.: Optical forces acting on a nanoparticle placed into an interference evanescent field. Opt. Commun. **275**, 409–420 (2007)
14. Tipler, P.A., Mosca, G.P.: Physics for Scientists and Engineers. W.H. Freeman and Company, New York (2007)
15. Burns, M.M., Fournier, J.M., Golovchenko, J.A.: Optical matter: Crystalization and binding in intense optical fields. Science **249**, 749–754 (1990)

---

[6] Note that due to the reduction in symmetry caused by the presence of the substrate, the rotational decomposition approach would no longer be appropriate, and the matrix would connect all eigenvalues $\mathbf{a}_{nm}$ with all other ones.

16. Abramowitz, M., Stegun, I.A.: Handbook of Mathematical Functions. Dover, Mineola, NY (1972)
17. Bobbert, P.A., Vlieger, J.: Light scattering by a sphere on a substrate. Phys. A **137**, 209–242 (1986)
18. Johnson, B.R.: Calculation of light scattering from a spherical particle on a surface by the multipole expansion method. J. Opt. Soc. Am. A **13**(2), 326–337 (1996)
19. Mackowski, D.W.: Exact solution for the scattering and absorption properties of sphere clusters on a plane surface. J. Quant. Spectrosc. Radiat. Transfer **109**, 770–788 (2008)

Chapter 4
Counter-Propagating Gaussian Beam Traps

"Currently, no theory has explained fully the occurrence of inhomogeneous particle spacing, both for a particle number dependency and a dependence on inter-array particle positions... and the spontaneous onset of oscillations observed in the dual beam trap" [1].

4.1 Introduction

In this chapter we will use numerical results from our Mie scattering model described in Chap. 2, as well as from a simpler heuristic model, to build up a detailed understanding of the mechanisms which lead to optical binding of a one-dimensional chain of trapped particles in a counter-propagating Gaussian beam trap. Section 4.4 discusses this on-axis trapping behaviour.

We will also discuss experimental observations of self-sustaining circulation within an optical trap, which in this case is *not* caused by beam misalignment but arises spontaneously from the physics of the optical binding interaction, despite the rotational symmetry of the perfectly-aligned Gaussian beam trap. The particles move in and out of the common beam axis as they circulate, in a driven harmonic motion. We will show that this type of behaviour is predicted by our rigorous Mie scattering model outlined in Chap. 2, and we will give a physical explanation for both static and dynamic off-axis trapped states. Section 4.5 discusses these off-axis trapping phenomena.

The dual-beam trap was possibly the first configuration in which 1D optically bound chains were observed [2, 3]. The inter-particle spacing decreases as the number of particles in the trap increases, and a slight anisotropy is observed in the chain whereby the inter-particle spacing towards the centre of the chain is smaller than the spacing at the edges of the chain [1], Fig. 4.5. Any comprehensive model of the optical interaction must be able to explain these results. As we discussed in Chap. 1, and as pointed out in [1], up until now no model has been able to achieve this.

We will show that despite the apparent similarities between the trapped chains in a counter-propagating beam trap and the trapped chains in counter-propagating Bessel beams [4], the binding mechanism is entirely different for a Gaussian beam trap, and is dominated by "radiation pressure" effects (the *scattering force* [5]). Now that it is clear that Mie scattering models can accurately reproduce experimental results first reported nearly seven years ago, the challenge is to interpret those results in easily-understood terms, and to distill out the key mechanisms

J. M. Taylor, *Optical Binding Phenomena: Observations and Mechanisms*,
Springer Theses, DOI: 10.1007/978-3-642-21195-9_4,
© Springer-Verlag Berlin Heidelberg 2011

Fig. 4.1 Schematic diagram
showing the experimental
setup for a simple counter-
propagating beam trap.
A Gaussian beam is expanded
and collimated, before being
split into two beams which
converge in a counter-
propagating geometry

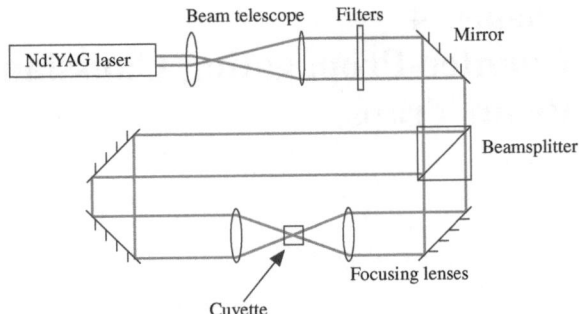

"hidden" within the complex model in order to develop a conceptual understanding of *why* optical binding and chain formation occurs. We will explain how
the trapped particles modify the beam shape to support stable chains of particles
with non-uniform inter-particle spacings, whose spacing decreases as the number
of particles in the trap increases.

4.2 Gaussian Beam Trap

Our experiment involves a trap formed from two orthogonally-polarized counter-
propagating Gaussian beams of vacuum wavelength 1064 nm, focused using a pair
of 50 mm focal length lenses to a beam waist radius of around 3 μm, with the
beam foci around 180 μm apart along their common axis. This is an experimental
configuration which is commonly referred to as a "counter-propagating optical
trap" or "dual-beam trap" [1–3, 6]. A schematic diagram of the simple experimental setup is shown in Fig. 4.1, and a close-up schematic of the trapping region
is shown in Fig. 4.2. This low numerical aperture configuration contrasts with
single-beam, high numerical aperture "optical tweezers" systems. The latter is
almost exclusively used for trapping and manipulation of a single particle per
beam, whereas "optical binding" of multiple particles is normally studied using
low numerical aperture systems like the one we discuss here [4, 7–10].

The most familiar form of trapping in such a trap involves the formation of a
chain of trapped particles, which can appear at first glance to be uniformly spaced,
but on closer inspection tends to display slight non-uniformities. An example of a
chain of trapped particles is shown in Fig. 4.3

4.3 Optical Binding Concepts, and Modeling

When light is incident on a particle, the particle scatters the light, producing a
secondary field which radiates in all directions. An important question to be
asked is: what governs the spacing of the particles—the forward-scattered field or

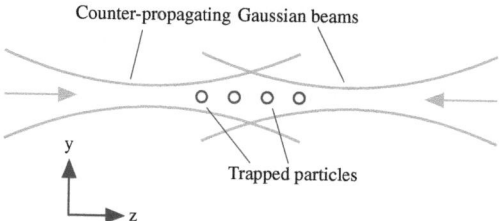

Fig. 4.2 Schematic diagram showing the detail of the trapping region of a counter-propagating beam trap. Two low numerical aperture Gaussian beams are focused to points a few hundred microns apart, with a common beam axis (the z axis). Microparticles are trapped in the region between the two focal points, traditionally forming a chain along the common beam axis

Fig. 4.3 Examples of stationary trapped particle chains of various lengths in our counter-propagating Gaussian beam trap. Here the particles are backlit with white light. Note that the particles are spaced closer together when the trapped chain is longer, and that particles at the centre of each chain are slightly closer together than particles at the edges of that chain

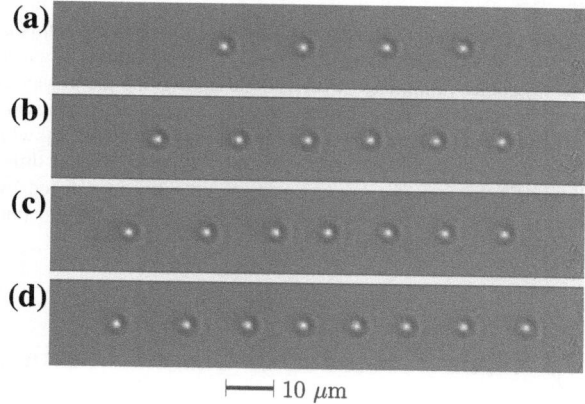

the back-scattered field? As has been observed by numerous authors [4, 6, 11], in the case of particles larger than the wavelength of the laser light, the interference of back-scattered light with the incident forward-propagating beam leads to a very large number of nearby trapped configurations, separated by either half a wavelength. In the Gaussian beam trapping geometry we are considering here, the typical particle spacings are considerably larger than the wavelength of the laser [2] (and indeed the particles may be several wavelengths in diameter). Consequently this binding mechanism, which is most significant for particles up to about half a wavelength in diameter, has little effect on the large-scale behavior of the trapped chains, which is determined largely by the forward-scattered light.

Because of this important distinction between the effects of forward- and back-scattering, all the models discussed in this chapter will explicitly treat the two incoherent counter-propagating beams separately: it is important that effects due to forward-scattering of one beam can be separated from back-scattering effects due to the other beam if the mechanisms are to be understood. We described a technique for "disabling" back-scattering in a theoretical calculation in Sect. 2.5.1.

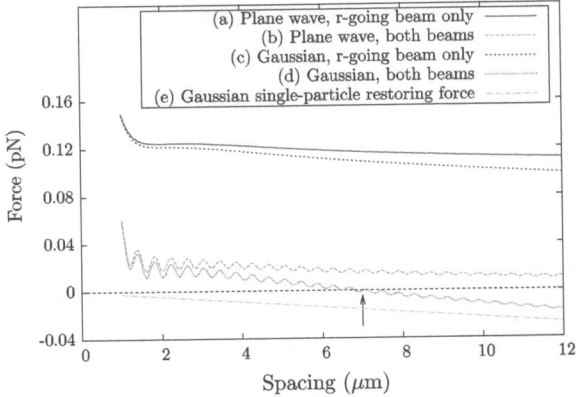

Fig. 4.4 Force on a 1 μm particle (size parameter ~ 3.9) in a two particle system, as a function of inter-particle spacing. This and subsequent graphs in this chapter were calculated using the theoretical model developed in Chap. 2. **a** Force on the downstream (right-hand) one of a pair of such particles illuminated with a single right-going plane wave. **b** as **a** but for a 25 mW Gaussian beam. **c** repulsive force between the pair of particles when two counter-propagating plane waves are used. **d** two counter-propagating Gaussian beams, showing the stable spacing of $\sim 8\,\mu$m. Note the modulation due to backscattered light, which has little effect on the general trend of the binding behavior. The broad harmonic potential introduced by the use of Gaussian beams has altered curve **c** to give curve **d**, which is outwardly similar but which has a stable inter-particle spacing at around 7 μm (marked with an arrow)

4.4 Chain Formation in Gaussian Beam Traps

4.4.1 Two Particles

In order to understand the binding mechanisms, consider first the intensity distribution downstream of a single 1.0 μm diameter particle illuminated by a single plane wave (Nd:YAG laser wavelength 806 nm in water; size parameter ~ 3.9). The dominant effect is a "focusing" of the light (in the limit of large particles we can consider the particle as a spherical lens within the framework of ray optics). Consider now the force on a second particle placed in this field. This is plotted in Fig. 4.4, which shows how two particles are stably bound in a Gaussian beam trap, but are not stably bound in counter-propagating plane waves. We can see that the force due to the light focused by the first particle causes the two particles to be repelled in the case of counter-propagating plane waves (Fig. 4.4c) . When we consider counter-propagating Gaussian beams (Fig. 4.4d), through symmetry there is no net force on the centre of mass of the particle pair, and the beams provide a broad background harmonic trapping potential. The particles will stabilize with a spacing which is largely determined by the balance of the repulsive force between the two particles and the harmonic trapping potential of the trap (Fig. 4.4d), as suggested in [2]. We emphasize that although we refer to a "focusing" of the light, we are far from the ray-optics regime, and it is not appropriate to use a ray-optics formula for the focal length, or to suggest that one particle will be bound "at the focus" formed by the other particle.

If we just consider the effects of a single beam, then in this experimental setup the contribution of the gradient force turns out to be only a small fraction of the total force on a particle. However, remember that with two counter-propagating beams a large part of the force exerted by one beam is balanced by the force due to the other beam. With some experimental parameters the contribution of the gradient force to the *net force* on a particle due to the two beams together can be non-negligible. Thus it is not really possible to state that either the gradient force or the scattering force will dominate under all circumstances, and the dominance will also to depend on the particle size.

4.4.2 Three or More Particles

A more interesting case than the two-particle case is to consider is that of a larger number N of trapped particles (indexed $i = 1$ to N), for which we intend to explain the three main properties of the particle chains:

- The fall in inter-particle spacing with N.
- The anisotropic particle spacing within the chain, with a smaller particle spacing near the middle of the chain.
- For some experimental parameters, chains are only supported up to a certain number of particles, beyond which the chains collapse.

There are a number of statements we can make about the behavior of the chains, based on symmetry considerations, with little or no assumptions on the nature of the inter-particle interactions:

1. Since the arrangement of the beams (two incoherent beams counter-propagating along the z axis) is symmetric about the $z = 0$ plane, the force on particle i due to one beam is equal and opposite to the force on particle $N - i + 1$ due to the other counter-propagating beam. This is true for any symmetric arrangement of particles, whether or not this is an equilibrium configuration. If f_i^+ (or f_i^-) is the force on particle i due to the beam propagating in the $+z$ (or $-z$) direction, then $f_i^+ = f_{N-i+1}^-$.
2. In addition, in equilibrium, there must be a net force of zero on each particle when the forces from the two beams are added together (i.e. $f_i^+ = f_i^-$), since by definition there must be no particle motion in equilibrium. Combining this with the previous requirement, we have $f_i^+ = f_{N-i+1}^+$. In other words, the forces on the particles in the chain due to *each individual beam* must be *symmetric* about the centre of the chain.

Fig. 4.5 A simple Born approximation model uses the intensity of a right-going Gaussian beam in the absence of particle 3 (whose location is indicated with an arrow in (c)) to determine the force f_3 on that particle. **a** shows the field for 3 particles exposed to a single Gaussian beam. **b** shows the field in the absence of particle 3 and **c** plots this field (all data in this figure was generated using a full Mie scattering model)

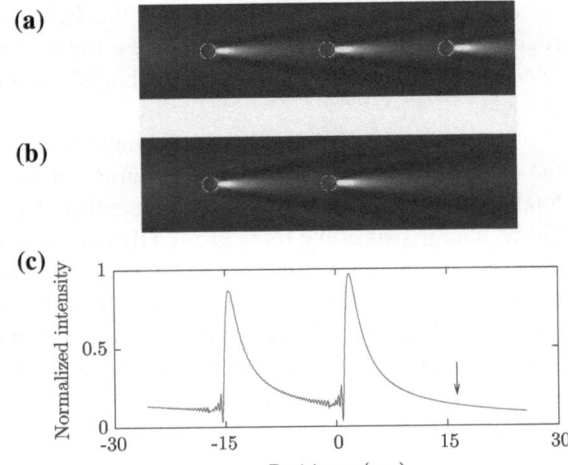

In addition to the above, we will make a number of simplifying hypotheses about the interaction mechanism between the particles. The following hypotheses have been observed to be approximately true in numerical experiments:

3. The particle spacings are determined by *forward*-scattering. As noted at the start of Sect. 4.3, this is a reasonable assumption for particles larger than the wavelength of light.
4. The force on a given particle i is a function of the light intensity $I_0^{(i)}$ which would be found be at that point *in the absence of* that given particle (the Born approximation; see Fig. 4.5).
5. The profile of the on-axis scattered intensity downstream of a given particle i has the form $I_0^{(i)} \times I(z - z_i)$, where $I(z - z_i)$ is a *fixed* downstream intensity profile which applies to any particle at any position. Consequently, the force on particle $i + 1$ has the form $I_0^{(i)} \times F(z - z_i)$ for a fixed downstream force profile $F(z - z_i)$ (per unit incident intensity). Although the downstream force profile $F(z - z_i)$ is *a function of* the intensity profile $I(z - z_i)$ (and its derivatives), the two are not necessarily *proportional*. If they were, that would be equivalent to stating that scattering forces dominate over gradient forces. We do not make that assumption in our model, though. As mentioned earlier, we find in practice that, while for some particle sizes the scattering force dominates overwhelmingly, the gradient force can also be significant in some cases.

These statements are already enough to explain why the inter-particle spacing at all points in the chain will *decrease* if an additional particle is added onto either end of the chain. The justification for this is as follows. The force pushing inwards on what is now particle 2 in the chain has been increased (it was previously just the force due to the unperturbed laser beam; it is now enhanced by the additional light focused onto it by particle 1). Assuming there are *some* losses along the length of the chain, then the force pushing outwards on what is now the last-but-one particle in the chain will also increase, but by *a smaller amount*. Hence whenever additional particles are added to the chain, the inner ones will be pushed inwards, and so the inter-particle spacing between any given pair of particles in the chain will decrease. Equilibrium is then restored because the closer inter-particle spacing enhances the transmission efficiency, thereby further increasing the repulsive force on the last particle in the chain.

In order to explain the non-uniform particle spacing, we propose a simple ansatz model for the force f_i as a function of z_i and z_{i-1}, as follows:

$$f_i(z_i, z_{i-1}) = e^{-\frac{z_i - z_{i-1}}{\alpha}} \times f_{i-1} + (\beta I_0 - \gamma z_i) \tag{4.1}$$

Here the first term represents assumption 5 (with the exponential decay providing a simple but not necessarily physically precise representation of the downstream force profile F referred to in assumption 5), and the second term represents a background intensity due to the laser field, which is decreasing with distance from the beam waist (for tunable parameters α, β and γ which depend on such things as the beam shape and the particle properties). We emphasize that the functional form of f_i, and its parameters, have simply been selected empirically to give a reasonable approximation to the observed inter-particle force. If a closer agreement with the Mie scattering model was desired, a "hybrid" model could be used, in which f_i is actually determined from the inter-particle forces for a pair of particles in a plane wave, calculated using Mie scattering theory. However, we have instead chosen to keep our model as elementary as possible.

4.4.3 Model Evaluation

Having made these assumptions, we can test the predictions of this very simple model against the definitive calculations of a rigorous Mie scattering model, and against established experimental observations. Figure 4.6 illustrates how this model predicts the force to vary along a chain of particles with a constant inter-particle spacing, and an example of how the forces calculated from a Mie scattering model vary along a similar chain with constant particle spacing. It can be seen from this plot that requirement 2 (a symmetric force profile) is not satisfied in either model: in the example shown, where the particles are spaced wider than their equilibrium position, the central particles in the chain feel the strongest overall compressive force and will move closer together the fastest. This then modifies the profile of the force plot, making it more symmetric. Conversely if the particles are all spaced closer than their equilibrium position then the strongest repulsive force will act on the end particles on the chain, again leading to a final configuration with non-uniform inter-particle spacings.

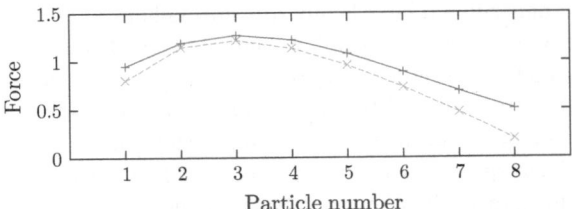

Fig. 4.6 Example of how the force (arbitrary units) on particle i in a chain of 8 particles due only to the right-going beam varies along the chain if the inter-particle spacing is constant. Mie scattering model (red, unbroken line) and our simple ansatz model (green, dotted line). Both plots show similar trends; neither is symmetric with respect to the centre of the chain. This means that, when both beams are considered, there will *not* be a net force of zero on a given particle, and so for this imposed uniform spacing the system will not be in equilibrium

Fig. 4.7 Example of how a force profile like those in Fig. 4.6 (arbitrary units) can be made symmetric by altering inter-particle spacings. Now, in contrast to Fig. 4.6 , when both beams are considered there will be a net force of zero on each particle, and so the system will be in equilibrium

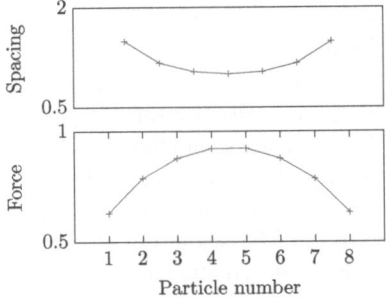

The natural next step is therefore to allow the particle spacings to vary along the length of the chain in our model, as they would do in real life in response to this repulsive force. As Fig. 4.7 shows, this approach allows a symmetric force profile to be produced, and leading to stable trapping of the chain with these slightly non-equilibrium spacings.

The first few particles in the chain act to focus the laser field onto the next particle in the chain, and hence initially the force rises sharply with particle index i. Particles towards the middle of the chain can be thought of as acting more like a (very inefficient) waveguide where the intensity is propagated from one particle to the next with some losses, which are compensated for by the re-focusing of additional background light. The intensity (and force) then drops again towards the end of the chain due to the increased particle spacings.

Finally, Fig. 4.8 shows how a chain above a critical length can collapse. The figure shows how the forces on the end particles in a short chain vary with inter-particle spacing for a particular set of experimental parameters (different to those used earlier, and carefully selected so the collapse occurs at an unusually short chain length). In this case a two-particle chain is supported, but if a third particle is added the chain will collapse until the spheres are in contact (an effect mentioned in [3]). As pointed out earlier, the compressive force on the first particle in the chain is greater if the chain has more particles in it (since that first

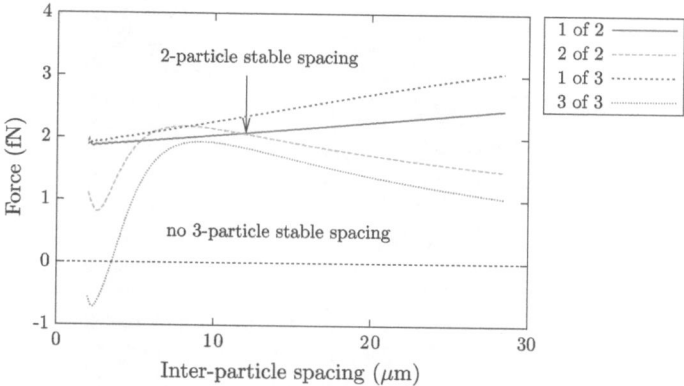

Fig. 4.8 Collapse of a longer chain, where a shorter chain would be stable. Parameters have been selected to give an extreme case where a two-particle chain is stable but a three-particle chain is not (parameters as [3], but with 1.9 μm diameter spheres). The curves "1 of 2" and "2 of 2" show the forces due to the right-going beam for the particles in a two-particle chain. Since the force on the second particle is greater than the force on the first particle over a range of around 5–12 μm inter-particle spacing, there is a stable trapped configuration with an inter-particle spacing of about 12 μm (indicated with an arrow) when net effect of both beams is considered. The curves "1 of 3" and "3 of 3" show the same forces on the end particles of a three-particle chain. Since there is no spacing for which the force on the third particle is greater than the force on the first particle, the chain will collapse. Even though there is some enhancement of the force on the third particle over what it would be in the absence of the other particles, this is not enough to overcome the compressive force due to the effect of the beam on the first particle

particle is closer to the beam waist). We argued earlier that this would cause the inter-particle spacing to fall until the inter-particle repulsive force was increased enough for the two forces to balance. However, for the particular parameters in this figure, at short inter-particle spacings there is in fact an *attractive* force between neighbouring particles, and so the three-particle chain collapses once the compressive forces have pushed the particles close enough to enter this regime. Our implicit assumption that the inter-particle light forces are repulsive (it was assumed that radiation pressure will dominate) has broken down; near-field gradient force effects have come into play, producing a net *attractive* force between the spheres at close ranges. There is no repulsive force to support the chain, and it collapses.

4.5 Off-Axis Trapped States

Figure 4.9 shows a series of frames from a video of particles circulating in our optical trap. 3.0 μm diameter silica beads are suspended in heavy water in the trapping region. It can be seen that the beads are circulating within the trap, with a period of approximately 10 s. The images are a composite of both a backlit

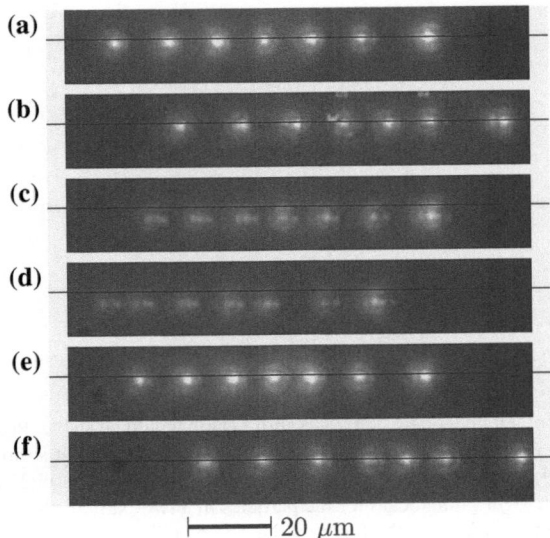

Fig. 4.9 Video frames from an experiment showing off-axis circulation of particles in an aligned Gaussian beam trap, taken over a 10 period. The red line indicates the approximate position of the beam axis. Frames **a-b** show the chain moving right, on the beam axis, **c-d** show the chain moving left, approximately 4 μm below the beam axis on the screen, and **e-f** show the chain moving right again, on the beam axis (all 6 frames have the same field of view). When on the beam axis the particles are exposed to a higher beam intensity, the particles move faster and scatter the light more strongly (the directly scattered laser light in the image is slightly offset due to chromatic aberrations in the optics and coherent scattering effects from the microspheres). The position of the beam axes was determined to under 1 μm by observation of the trajectories of isolated particles exposed to a single beam at a time. Similarly, the beam alignment in the x direction (into the page) was determined to within 2 μm based on the change in focus of the particle images as a function of x coordinate (but, as implied by the results in [1], we would not expect misalignment along the x axis to lead to circulation in the yz plane anyway)

white-light transmission image of the particles, and direct imaging of the laser light scattered from the particles (the imaged scattered light is slightly offset relative to the white light image due to chromatic aberrations in the optics and coherent scattering effects from the microspheres). The white light illumination is purely to aid with the viewing of the particles, and is nowhere near intense enough to affect the inter-particle interactions. It can be seen from the change in brightness of the scattered light that the particles are on the beam axis as they move to the right, but are on the edge of the Gaussian beam, where they are exposed to a weaker light intensity as they move to the left.

In nearly all experimental and theoretical papers published on optical binding, the particles are confined to the beam axes [1, 3, 6, 12, 13], leading to the one-dimensional problem we have considered in Sect. 4.4 In Ref. [1], circulating modes were reported, but these were caused by a deliberate misalignment of the trapping beams.

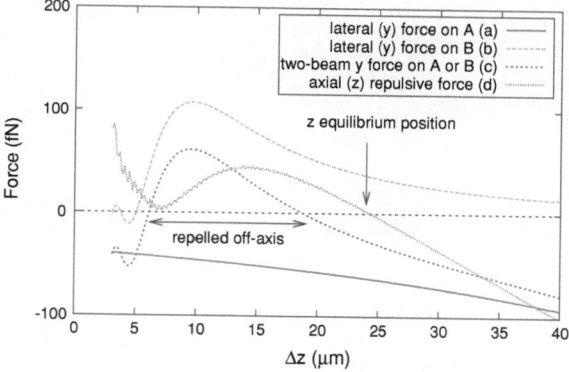

Fig. 4.10 Forces acting on a pair of trapped 3 μm diameter silica particles as a function of particle spacing Δz along the z axis. The particles are very slightly offset from the beam axis (by approximately 10 nm). Positive forces represent repulsion. If we first just consider the rightgoing beam, particle A is drawn on axis (**a**), but B is pushed off axis for a large range of particle spacings (**b**). When the effects of both beams are considered (**c**), both particles are pushed off axis for z spacings between 6 and 19 μm. In this case that range does not coincide with the equilibrium particle spacing in z (**d**)

In this section we will discuss the physical mechanisms behind the off-axis trapping behaviour we have observed. We have reproduced behaviour similar to that shown in Fig. 4.9 using our Mie scattering model, as we will see later in Fig. 4.16 , but we will begin by considering Mie scattering calculations for smaller numbers of particles in order to understand the mechanisms underlying this behaviour.

4.5.1 Lateral Forces on Two Trapped Particles

We modelled a counter-propagating beam trap like that described in Sect. 4.2 but containing only two particles. Figure 4.10a plots the lateral force on the first of the two particles and Fig. 4.10b the force on the second of the particles, in both cases just considering the effects of right-going of the two trapping beams, when both particles are slightly offset from the beam axis. Because, under the experimental parameters for those plots, the particles interact with the field largely through forward scattering (as discussed in [10]), the first particle behaves very similarly to how an isolated single particle would behave in the trap. Through the gradient force, it is pulled back onto the beam axis from its initial small offset. In contrast, the graph shows that the second particle is pulled *further* off axis by the focusing effects of the first particle on the field. As shown in the intensity map in Fig. 4.11, the "plume" of light focused by the first particle is angled slightly off-axis due to the diverging nature of the incident beam at this point and the lens-like behaviour

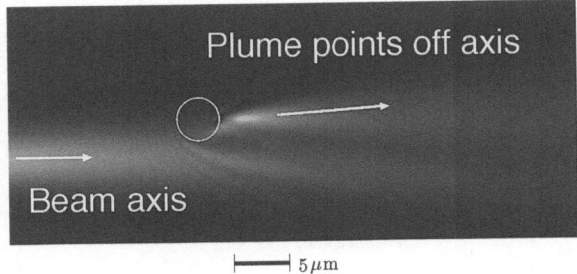

Fig. 4.11 Field intensity around a single off-axis particle, showing the "plume" of light focused by the particle (in this case a 3 μm diameter silica sphere). Due to the diverging nature of the beam, this is angled slightly off-axis, and so a second particle will be drawn *even further* away from the axis through the gradient force

of the sphere. Through the same gradient force, this causes the second particle to be pulled further off axis.

When the effects of both counter-propagating beams are considered, there is a range of inter-particle spacings, for which the lateral (y) force (which through symmetry is the same for both particles) acts to repel the particles from the beam axis. In the case of Fig. 4.10c this range is between 6 and 19 μm. For this range of inter-particle spacings the particles are in a state of unstable equilibrium on the beam axis, and a small perturbation (due for example to Brownian motion) will be amplified and cause the particles to move away from the beam axis. However, Fig. 4.10d shows the z force (force parallel to the beam axes) as a function of inter-particle spacing, demonstrating that in this particular scenario, *when the particles are at their equilibrium separation* in the i direction the particles are stably trapped on the beam axis, as is conventionally the case in optical trapping, and stable off-axis trapping does not occur in this two-particle case.

4.5.2 Lateral Forces on Three Particles

In the previous section we found a lack of coincidence between y repulsion and the equilibrium spacing in z. This applies over all the parameter space we have explored for two particles, and there is a reason for that. As stated earlier, the y repulsion relies on the bright "focal plume" of one particle drawing the other particle off axis. However it is this same bright focal plume which acts to repel the particles in z, pushing them apart to a greater equilibrium spacing [10]. However, if we introduce more particles into the trap, the particles are forced closer together [1, 2, 10], and it is possible for the y repulsion mechanism to act at the equilibrium chain spacing. Figure 4.12 shows the force on a chain of three trapped particles, for which the net y force on the chain will act to push it off axis.

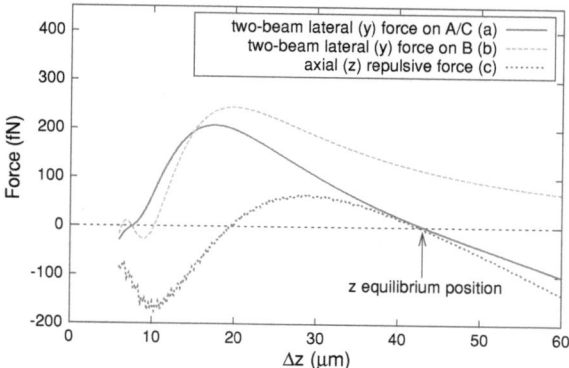

Fig. 4.12 Forces acting on three trapped 3 μm diameter silica particles (labeled A, B and C from left to right) as a function of particle spacing in z, At the equilibrium inter-particle spacing (determined from curve (**c**)), the end particles feel very little lateral force (**a**), but the central particle is pushed off axis (**b**). The net result is that the three particle chain is pushed off axis

Figure 4.13a shows the trajectory of three particles which are initially positioned very close to the beam axis. It can be seen that the particles settle into a stable trapped configuration away from the beam axis (arrows represent stable particle positions). The state has not recovered from the tiny initial perturbation, but has switched from its initial unstable equilibrium on axis to a completely different state which does not conform to the symmetry of the trap.

Figure 4.13b shows a similar trajectory, but for particles with a slightly higher refractive index, which causes them to interact more strongly with the beams. The particles are pushed off axis as before, but instead of settling into an entirely stationary condition, their trajectory stabilizes into a closed orbit. Such *limit cycle* behaviour was previously reported by Ng et al. [8] for clusters of particles trapped in the plane perpendicular to coherent counter-propagating plane waves, but in our experiment the motion is in a plane parallel to the beam, and does not conform to the symmetry of the trap. We emphasize that the particles are in a heavily over-damped regime in which free harmonic oscillation cannot be supported. It is the continual input of energy into the system by the trapping beams which drives this harmonic motion.

As the refractive index is increased still further, the scale of the limit cycles grows, until a macroscopic circulation within the trap develops (as shown in Fig. 4.13c). The particles circulate in a figure-of-eight pattern around the trap. This motion cannot be described in terms of particles moving subject to a single con-servative potential—indeed in the over-damped case, sustained motion cannot occur in such a model—but we can describe the competing effects which give rise to the macroscopic circulation. A general feature of the motion is that movement in the y direction, perpendicular to the beam axis, is much more rapid than movement in the z direction, parallel to the beam axis. Broadly speaking, this is because there is a strong gradient to the laser field in the y direction, leading to

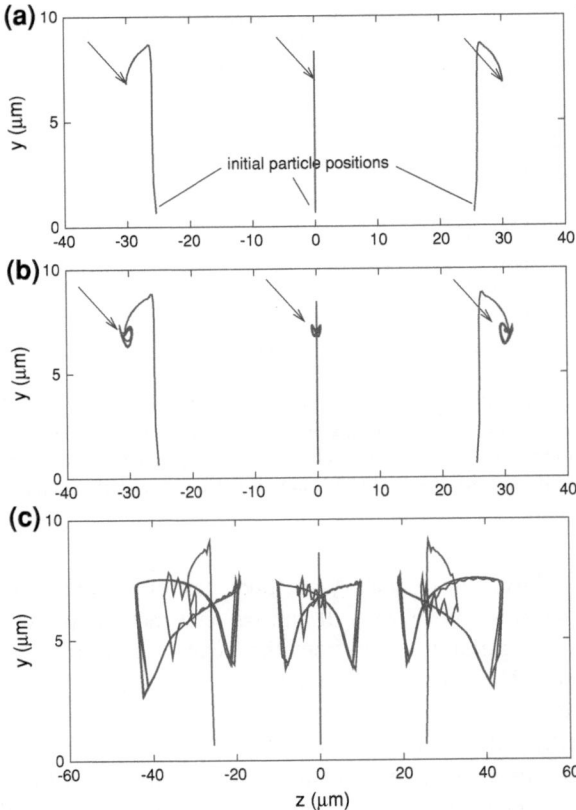

Fig. 4.13 Simulated trajectories of three 1.9 μm diameter silica particles in the trap (refractive index of surrounding medium is 1.35). The particles start close to the beam axis at the start of the simulation. In **a** they are rapidly pushed off-axis and then stabilize at the positions marked by the arrows. The particles are stationary, but are positioned away from the beam axis rather than being trapped on axis in accordance with the symmetry of the trap. In **b** where the particles have a slightly larger refractive index, the particles attain similar positions but are not completely stationary. They follow small oscillatory trajectories. In **c** where the refractive indices are still higher, the particles eventually stabilize into a wide-ranging closed trajectories in which they circulate around the trap indefinitely. If alternatively the particles are positioned at a small negative y coordinate, instead of a small positive one as shown in these figures, then the resultant trajectories are, as expected, identical mirror images of those shown here, reflected in the z axis

strong gradient forces in that direction. The net forces in the z direction are weaker because the force due to each of the two beams act in opposing directions, and it is only the *difference* between those forces (of similar magnitude to each other) that leads to motion of the particles in the z direction (see [10] for further discussion).

If the particles begin in the centre of the trap, on the beam axis, then as we have seen they will be repelled from the beam axis (in response to a small initial perturbation from the unstable on-axis position, caused by Brownian motion),

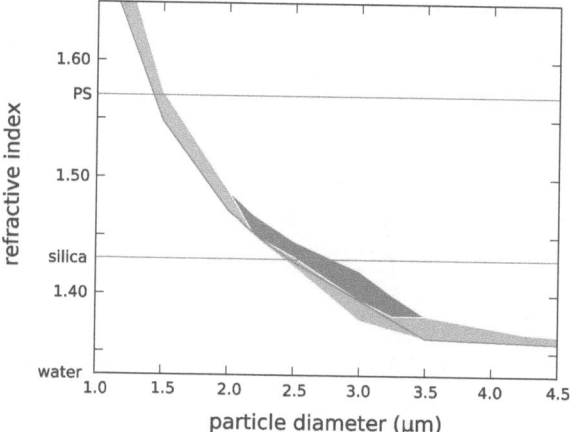

Fig. 4.14 Simulation results showing the range of parameters for which off-axis trapping is seen for three particles. The lightly shaded region indicates stationary off-axis trapping. For all particle sizes investigated (between 1 and 5 μm in diameter) it is possible to see off-axis trapping if the refractive index is selected appropriately. In the range of diameters 2–3.5 μm, dynamic off-axis states are seen (the heavily shaded region). Below the red line, stationary on-axis trapping is supported, and it can be seen that there is a region of bistability where both on- and off-axis stationary trapping is supported (the red line overlaps with the lightly shaded region). The data in this plot was generated for a beam waist diameter of 5 μm, and beam foci 180 μm apart. The results shown ignore Brownian motion; for sufficiently weak beam powers its effects will be to reduce the extent of the various trapping regimes

stabilizing at some radius away from the axis. For sufficiently large particle size and refractive index, though, this configuration is in turn unstable in z. If one of the end particles is perturbed slightly away from the centre of the chain, that motion will be amplified, and the centre of mass of the chain will move in the direction of that initial perturbation. Thus the chain moves away from the $z = 0$ position. This motion continues until such a point as the inter-particle spacings and the position of the particles relative to the beam waists has changed enough that our analysis of Fig. 4.12 , which assumes the centre of mass of the particles is at $z = 0$, no longer applies, and the particles are no longer repelled from the beam axis. This means that the particles are rapidly pulled back on–axis by the gradient force, at which point they are pushed back towards the centre of the trap ($z = 0$) by the imbalance of the radiation pressure experienced from the two beams. The process then repeats and the cycle continues.

Figure 4.14 shows simulated results giving the range of particle diameters and refractive indices for which on- and off-axis trapping is possible for three particles. That plot does not indicate how probable it is that these states will be produced by the natural "wandering" of particles into the trapping region one by one. Figure 4.15 gives an indication of the stability of a stationary off-axis trapped state, showing the range of starting positions which end up in the off-axis state.

In the results we have shown up to now the circulation has been shown in the yz plane, but it might appear that there is no preferred orientation of the plane of

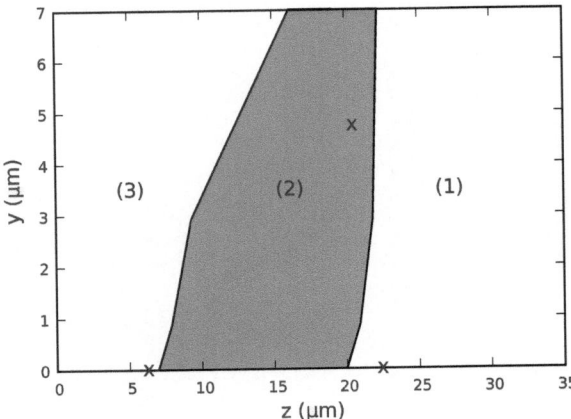

Fig. 4.15 Simulation results showing what happens when three equally-spaced particles are placed at positions (y,-z), (y,0) and (y,z) and then allowed to move under the effects of the trapping beams. The particles are 3 μm in diameter and the refractive index ratio between the particles and the surrounding medium is approximately 1.044 (which can be achieved by immersing silica particles in a weak sucrose solution). The beam parameters are the same as in Fig. 4.14 . Starting positions in region **1** end up in the on-axis trapping configuration indicated by a cross (z = 22.4 μm). Starting positions in the shaded region, **2** end up in the stationary off-axis trapping configuration indicated by a cross (y = 4.7 μm, z = 20.4 μm). The system in fact supports a second, closer on-axis trapped state, which particles starting in region, **3** will end up in. If alternatively the refractive index ratio is increased to 1.05 then, over the *whole range* of initial positions shown in the plot, the particles will end up in a circulating off-axis trapped configuration

circulation (for example circulation could equally well occur in the *xz* plane depending on the initial conditions. In a real experiment a preferred direction tends to be introduced by residual convection currents in the surrounding liquid. It is for this reason that the circulation in Fig. 4.9 takes place in the plane of the video screen. Even in a numerical simulation, without any convective effects, it turns out that there is a slightly preferred direction introduced by the polarization of the beams: although circulation will begin in a plane determined by the initial conditions, in the absence of external influences the plane of circulation will gradually rotate over around ten periods of oscillation until the plane is perpendicular to the polarization of the beams.

The description we have given in this section explains the physical mechanisms behind the circulating modes in the trap, where for ease of understanding we have chosen the simplest case of three trapped particles. We note that parameters such as the particle properties and the beam waist size were carefully selected to display this symmetry breaking (repulsion from the beam axis) in a short three particle chain. For the majority of parameters, the effect is not strong enough to repel the three particles from the beam axis. However, as larger numbers of particles join the chain, the repulsion is enhanced because the particles are pushed closer together as more of them are confined within the fixed trapping region (see [2, 14] for experimental results and [10] for a theoretical discussion of this effect), and a

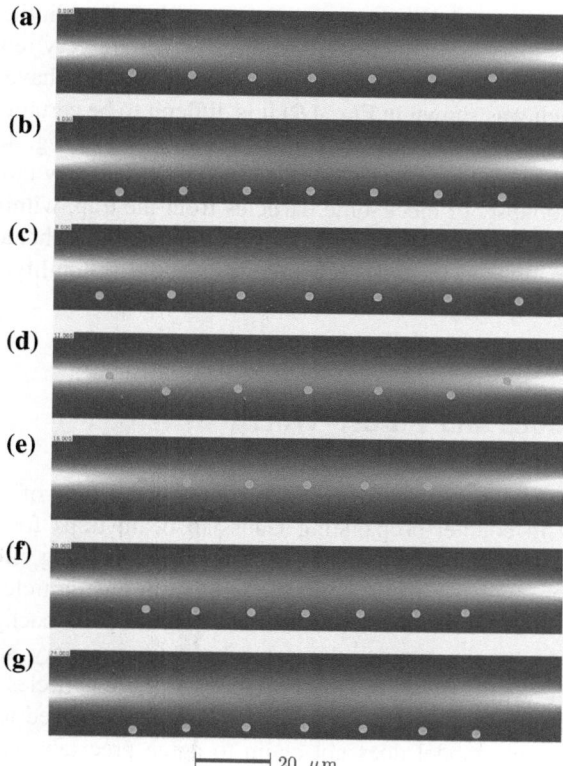

Fig. 4.16 Simulated video frames at four second intervals (for a beam power of 25 mW) showing off-axis circulation with a larger number (seven) of 1.9 μm polystyrene beads. **a-c** show the chain expanding off axis. **d** and **e** show the chain moving on axis again and being compressed. **f** and **g** show the chain pushed back off axis. In addition to showing that circulation occurs in simulations with large numbers of particles there are two interesting points about this example. Firstly, the motion is symmetrical about the plane $z = 0$ (contrary to the case discussed earlier, this chain is stable against perturbations in z). Secondly, in this case the mode is not completely stable, and after a few periods of circulation the chain collapses until the particles are touching (which is an effect often observed in experiments as well). Behaviour such as that shown in this simulation may be behind the "breathing" modes reported in [2]

critical point will be reached at which the symmetry of the trapped state can be broken. In the case of our experiment this occurred for chains of between 7 and 10 particles (as illustrated in the video frames shown in Fig. 4.9).

4.5.3 Larger Numbers of Particles

Finally, to demonstrate that our theoretical model can be extended to larger numbers of particles, Fig. 4.16 shows a simulation of 7 trapped polystyrene particles exhibiting circulation around the trap. As described in the figure caption, this simulation shows some interesting features which are very reminiscent of the experimental results reported in [2]. In the experiments we have performed (an example of which was shown in Fig. 4.9) it is difficult to be certain whether we are observing truly stable circulation, such as that simulated in Fig. 4.13, or unstable circulation, such as that simulated in Fig. 4.16. The circulatory modes we observe tend to either collapse, or eject some particles from the trap, within a few periods of oscillation, but it is not possible to determine for certain whether this is due to external perturbations to the system or due to inherent instability in the observed modes of oscillation.

4.6 Conclusions and Future Work

We have explained the mechanisms behind the formation of optically bound particle chains in counter-propagating Gaussian beam traps for particles larger than the wavelength of the trapping laser. The optical binding effect results from the balancing of repulsive effects from the light from one particle incident on the next particle in the chain and compressive effects due to the background trapping potential formed by the beams.[1] Here we have used a very simple model to successfully explain the trends of closer spacings as more particles are added to the chain, and of closer spacings in the centre of a chain compared to near its edges.

While our simple model does not claim to agree precisely with experimental results and with the theoretical gold standard of Mie scattering calculations which we have also used (and has a number of parameters which must be tuned by hand), there is good qualitative agreement between them across a range of model parameters. From this we can conclude that, while there is some influence from more sophisticated effects which can only be encapsulated in a full vector model based on rigorous solution of Maxwell's equations (such as Mie scattering), many of the properties of the trapped particle chains can be understood in terms of a simple scalar model. This model can offer strong conceptual insights into the physical mechanisms which lead to the observed behavior, which had not previously been fully explained. As well as explaining how the inter-particle spacings are regulated, it explains the trend for closer spacings with larger N, and the wider spacing close to either end of the chain. It is *only* through a simple model such as the one we have presented that the various complex effects in the experiment can be decoupled in order to understand *why* optical binding occurs under these experimental conditions.

---

[1] It could in fact be argued that this effect is not optical "binding" in the strict sense shown in early experiments [7], since the interaction here is largely a repulsive one, with stable chains only being formed due to the background harmonic potential of the trap.

We also emphasize again that the mechanism we have discussed here is very different from that described in [4] for a counter-propagating Bessel beam trap. There the optical binding, and the inter-particle separations, was determined by coherent interference effects between the background laser field and the light forward-scattered by the particles.

In addition to studying on-axis trapping, we have shown experimental and theoretical examples of spontaneous symmetry breaking leading to asymmetric circulatory motion within the trap. We have shown that this is predicted by a Mie scattering model, and we have explained the mechanisms behind this behaviour in simple physical terms. Critical to this effect is the concept that the presence of trapped microspheres within the trap strongly modifies the electromagnetic field within the trap, such that the evolution of the system is governed predominately by the light-mediated inter-particle interactions rather than the background trapping potential.

Our discussion has focused on a low numerical aperture configuration which is standard in optical binding experiments, but there is potential for further investigation into whether these effects have an impact in high numerical aperture configurations with multiple trapped particles, such as that discussed in [15].

We have demonstrated that off-axis trapping can occur with as few as three particles and this shows that, even with small numbers of trapped particles, the full inter-particle interactions must be considered in order to be able to correctly predict the behaviour of the system.

References

1. Kawano, M., Blakely, J.T., Gordon, R., Sinton, D.: Theory of dielectric micro-sphere dynamics in a dual-beam optical trap. Opt. Express **16**, 9306–9317 (2008)
2. Tatarkova, S.A., Carruthers, A.E., Dholakia, K.: One-dimensional optically bound arrays of microscopic particles. Phys. Rev. Lett. **89**(28), 283901 (2002)
3. Singer, W., Frick, M., Bernet, S., Ritsch-Marte, M.: Self-organized array of regularly spaced microbeads in a fiber-optical trap. J. Opt. Soc. Am. B. **20**(7), 1568–1574 (2003)
4. Karasek, V., Brzobohaty, O., Zemanek, P.: Longitudinal optical binding of several spherical particles studied by the coupled dipole method. J. Opt. A. **11**, 034009 (2009)
5. Maria, Dienerowitz., Michael, Mazilu., Kishan Dholakia.: Optical manipulation of nanoparticles: A review. J. Nanophotonics **2**, 021875 (2008)
6. McGloin, D., Carruthers, A.E., Dholakia, K., Wright, E.M.: Optically bound microscopic particles in one dimension. Phys. Rev. E. **69**, 021403 (2004)
7. Burns, M.M., Jean-Marc , Fournier., Golovchenko, J.A.: Optical binding. Phys. Rev. Lett. **63**(12), 1233–1236 (1989)
8. Jack, N.g., Lin, Z.F., Chan, C.T., Ping, Sheng.: Photonic clusters formed by dielectric microspheres: Numerical simulations. Phys. Rev. B. **72**, 085130 (2005)
9. Metzger, N.K., Dholakia, K., Wright, E.M.: Observation of bistability and hysteresis in optical binding of two dielectric spheres. Phys. Rev. Lett. **96**, 068102 (2006)
10. Taylor, J.M., Love, G.D.: Optical binding mechanisms: A conceptual model for Gaussian beam traps. Opt. Express **17**(17), 15381–15389 (2009)
11. Tomás , Čižmár., Věra , Kollárová., Zdeněk, Bouchal., Zemánek, Pavel.: Sub-micron particle organization by self-imaging of non-diffracting beams. New J. Phys. **8**, 43 (2006)

12. Metzger, N.K., Wright, E.M., Sibbett, W., Dholakia, K.: Visualization of optical binding of microparticles using a femtosecond fiber optical trap. Opt. Express **14**(8), 3677–3687 (2006)
13. Metzger, N.K., Wright, E.M., Dholakia, K.: Theory and simulation of the bistable behaviour of optically bound particles in the Mie size regime. New J. Phys. **8**, 139 (2006)
14. Gordon, R., Kawano, M., Blakely, J.T., Sinton, D.: Optohydrodynamic theory of particles in a dual-beam optical trap. Phys. Rev. B. **77**, 245125 (2008)
15. Roichman, Y., Grier, D.G.: Three-dimensional holographic ring traps. Proc. SPIE. **6483**, 64830F (2007)

Chapter 5
Conclusions

5.1 Discussion

A common theme throughout this thesis has been the unpredictable nature of optical binding interactions between multiple particles. We observe "emergent" phenomena when multiple particles are trapped together, which cannot easily be predicted by considering each particle in isolation. Specifically we considered the following phenomena:

- Interaction strength: We saw in Sect. 3.4 that for a two-dimensional cluster the strength of the optical binding force on a given particle will scale with the radius of the cluster. This has a number of important implications:

  - The optical binding force will increase until it has changed the background laser field beyond all recognition. This has important implications for holographically-generated "optical landscapes" (see for example [1, 2]). It is all very well holographically generating the desired optical landscape, but the effects this landscape will have on the trapping of large numbers of particles cannot easily be predicted. It is not sufficient simply to consider the interaction of a single particle with that landscape, and then assume that large numbers of particles will interact in the same way.
  - This increasing force, and the underlying interference effects, will tend to limit the maximum supported size of clusters. It has been observed in experiments and in theoretical investigations that there appears to be an upper limit on the size of stable clusters in a given configuration [3–5]. Our own investigations have encountered the same effects.
  - An interesting side-effect of the scaling of the optical binding interaction is that we cannot just perform a *local* analysis of a two-dimensional structure- the cluster must be considered as a whole. The magnitude of the effects of spheres lying in a shell at a large radius R, between R and $R + \Delta r$ will be the

J. M. Taylor, *Optical Binding Phenomena: Observations and Mechanisms*,
Springer Theses, DOI: 10.1007/978-3-642-21195-9_5,
© Springer-Verlag Berlin Heidelberg 2011

same as the effect of spheres lying in a closer shell between r and $r + \Delta r$. Furthermore, it is *not* appropriate to consider a cluster in the limit of infinite size, since at this point the forces on each particle in the cluster will tend to infinity.

This growth in the interaction strength, and the observation that there appears to be an upper limit on the size of stable clusters, unfortunately appears to suggest that very large-scale applications for self-organizing structures, such as the space-based telescope proposed in [6], will present challenges that may be unsurmountable.

- Fringe affinity: We saw in Sect. 3.5 that sudden, fundamental changes can occur as the number of trapped particles increases. Specifically, we found that with a critical number of particles trapped parallel to a series of interference fringes, the particles can switch from being trapped on a fringe minimum to a fringe maximum.
- Lateral stability in Gaussian beam traps: We saw in Sect. 4.5 that beyond a critical number of particles (which may be as few as three), it is possible for an on-axis trapped chain to be in unstable equilibrium: the symmetry will be broken and the particles will make a transition to a different state which does not reflect the axial symmetry of the trapping beams. Not only do we find off-axis trapped states, but these states may not even be stationary states. Under certain conditions these states can form stable or partially-stable circulating modes.

These cases all illustrate how important it is to have a comprehensive understanding of the inter-particle interaction. If only single-particle interactions had been considered, it would not be possible to predict or explain the phenomena we have described. As the interest in multiple-particle trapping and manipulation continues, the relevance of optical binding effects can only increase.

5.2 Summary of Achievements

The key results of this thesis are:

- Development of a sophisticated computer model based on Mie scattering theory. This has drawn together a range of theoretical results published in the literature, and has resulted in a model which is many tens of times faster than any available alternatives.
- New theoretical results for Mie scattering theory. These include derivations of the beam expansion coefficients for Bessel and Gaussian beams in Sects. 2.4.4 and 2.4.5, and the derivations of the expressions for the gradient potential (Sect. 2.6.2) and pressure inside a liquid droplet (Appendix C.2).
- Interpretation of evanescent wave trapping results. We have shown that we can reproduce some of the structures observed experimentally in evanescent wave

traps. We have made experimentally testable predictions about the fringe affinity of these structures. We have highlighted the importance of the interplay between optical binding forces and physical close-packing restrictions between adjacent particles.

- Novel counter-propagating beam trap results. We have reported experimental results which show particle behaviour that is not governed by the underlying symmetry of the trapping beams. We have shown driven harmonic motion even in the absence of any measurable beam misalignment, in both experiment and numerical simulations.

- Mechanism for chain formation in counter-propagating Gaussian beam traps. We have explained in simple physical terms how particle chains in Gaussian beam traps are formed and regulated. We have explained the non-uniformity of the inter-particle spacing as a function of particle number and position within the chain.

- Mechanism for off-axis trapping and circulation in counter-propagating Gaussian beam traps. We have shown how particle chains can lose confinement in the lateral direction and move off axis. We have outlined an explanation for how this leads to off-axis trapping and circulating modes in the trap.

5.3 Future Directions

The findings reported in this thesis leave many open questions which deserve further attention. Some areas worth of closer investigation include:

- Evanescent wave trap

  - Experimental demonstration of fringe affinity in particle clusters based on the theoretical predictions made in Chap. 3.
  - Further investigation of 1D and 2D cluster formation including the intriguing results shown in Fig. 3.11 involving two different particle sizes.
  - Effect of the substrate surface on optical binding in an evanescent wave trap. To our knowledge this aspect of the optical binding interaction has not been investigated, and the substrate may well prove to have a significant influence on the physics of the binding interaction.
  - Other aspects listed in Sect. 3.7.

- Gaussian beam trap

  - Further analysis of the off-axis circulating modes, with the aim of developing a more detailed understanding of the processes involved.
  - Theoretical investigation of the processes involved in chain collapse. We have observed ejection of some or all particles from the trap during collapse, and high speed "corkscrew" and "whipping" effects after particles have coalesced within the trap.

- High-N.A. optical binding. To our knowledge there has been no investigation of optical binding effects in high N.A. optical tweezers experiments. In Sect. 1.1.1 we listed a range of optical tweezers applications, some of which may display optical binding effects and some of which are unlikely to. Time-sharing traps should not display optical binding effects since only one particle is illuminated at a time (see further discussion in Sect. 3.4). Holographic and vortex-like traps, however, can trap multiple particles in close vicinity to each other, with all the particles scattering mutually coherent light. In this case we would expect that each particle would interact with the scattered light from other nearby spheres, and this light-mediated inter-particle interactions should be significant. This is an area which deserves closer attention.
- Comparison between optical binding regimes. In this thesis we have investigated optical binding in a Gaussian beam trap, and we have also investigated the very different binding mechanisms in Bessel and evanescent wave traps. Those additional findings (not reported here) are in agreement with findings reported recently for Bessel beam traps [7]. It would be nice to consider all three binding regimes together, in terms of the similarities and differences between their respective mechanisms.
- Release of computer code. As stated earlier, our code has considerably higher capabilities that anything currently publicly-available and, given sufficient time to tidy it into an easily-used form, it would be nice to make it more widely available.

References

1. Liesener, J., Reicherter, M., Haist, T., Tiziani, H.J.: Multi-functional op- tical tweezers using computer-generated holograms. Opt. Commun. **185**, 77–82 (2000)
2. Curtis, J.E., Koss, B.A., Grier, D.G.: Dynamic holographic optical tweezers. Opt. Commun. **207**, 169–175 (2002)
3. Fournier, J.-M., Boer, G., Delacrétaz, G., Jacquot, P., Rohner, J., Salathé, R.P.: Building optical matter with binding and trapping forces. Proceedings of SPIE. **5514**, 309–317 (2004)
4. Karásek, V., Čižmár, T., Brzobohatý, O., Zemánek, P., Garcés-Chávez, V., Dholakia, K.: Long-range one-dimensional longitudinal optical binding. Phys. Rev. Lett. **101**, 143601 (2008)
5. Hang, Z.H., Ng, J., Chan, C.T.: Stability of extended structures stabilized by light as governed by the competition of two length scales. Phys. Rev. A. **77**, 063838 (2008)
6. Labeyrie, A., Fournier, J.-M., Stachnik, R.: Laser-trapped mirrors in space: Steps towards laboratory testing. Proceedings of SPIE. **5514**, 365–370 (2004)
7. Karásek, V., Brzobohatý, O., Zemánek, P.: Longitudinal optical binding of several spherical particles studied by the coupled dipole method. J. Opt. A. **11**, 034009 (2009)

Curriculum Vitae

Jonathan Taylor

Education	
2005–2009	Durham University
	Ph.D. in Physics, awarded November 2009
	(supervisor Dr Gordon Love)
2001–2005	University College, Oxford University
	1st class MPhys degree in Physics
1995–2000	Bryanston School, Blandford, Dorset

Current Research and Other Activities

Senior Research Associate, Durham University, June 2009 to Present

Since completing his Ph.D. Jonathan has mainly worked on two EPSRC funded research projects, *"Beating Hearts at High Resolution: Adaptive High Resolution Selective Plane Illumination Microscopy"* and *"Optical Control of Emulsion Drops for Nanofluidics and Nanofabrication"*.

Jonathan has also translated a number of physics and engineering texts from French into English for publication, and in his spare time he is a keen climber and a member of a mountain rescue team.

J. M. Taylor, *Optical Binding Phenomena: Observations and Mechanisms*,
Springer Theses, DOI: 10.1007/978-3-642-21195-9,
© Springer-Verlag Berlin Heidelberg 2011

List of Publications and Conference Presentations

Journal papers

1. "Realtime Optical Gating for 3D Heart Imaging", **J. M. Taylor**, C. D. Saunter, B. Chaudhry, D. Henderson and G. D. Love, J. M. Girkin. Submitted to Journal of Biomedical Optics (2011).
2. "Nanofluidic Networks Created and Controlled by Light". D. Woods, C. D. Mellor, **J. M. Taylor**, C. D. Bain and A. D. Ward. Soft Matter **7** 2517 (2011).
3. "Spontaneous Symmetry Breaking and Circulation by Microparticle Chains in Gaussian Beam Traps". **J. M. Taylor** and G. D. Love. Physical Review A **80** 053808 (2009).
4. "Optical Binding Mechanisms: A Conceptual Model for Gaussian Beam Traps". **J. M. Taylor** and G. D. Love. Optics Express **17** 15381–15389 (2009).
5. "Multipole Expansion of Bessel and Gaussian Beams for Mie Scattering Calculations". **J. M. Taylor** and G. D. Love. Journal of the Optical Society of America A **26** 278–282 (2009).
6. "Emergent Properties in Optically Bound Matter". **J. M. Taylor**, L. Y. Wong, C. D. Bain and G. D. Love. Optics Express **16** 6921–6929 (2008).

Conference Presentations

1. Presented paper: "Adaptive selective plane illumination microscope with image synchronization", SPIE Photonics West, San Francisco, USA, January 2011.
2. Lecture: "Zebrafish Heart Imaging with Selective Plane Illumination Microscopy", Workshop on Imaging for Life Science Research, Durham, January 2011.
3. Presented paper: "SPIM in cardiac development: heartbeat synchronization", 2nd Light Sheet Fluorescence Microscopy workshop, Dublin, September 2010.
4. Contributed to conference paper: "Directed assembly of optically bound matter", Photon 10, Southampton, August 2010.
5. Contributed to conference paper: "Deep diffractive liquid crystal lenses", Photon 10, Southampton, August 2010.
6. Invited paper (in conjunction with supervisor Gordon Love): "On- and off-axis binding and induced circulation in counter-propagating beam traps", SPIE Photonics West, San Francisco, USA, January 2010.
7. Presented paper: "Optical binding mechanisms", CLEO 09 conference, Munich, Germany, June 2009.
8. Presented paper: "Emergent behaviour of optically bound microparticles in laser beam traps", Photon 08 conference, Heriot-Watt University, August 2008.
9. Presented paper: "Full Mie Scattering Model of Optically Bound Particles in Evanescent Waves", Progress In Electromagnetics Research Symposium in Hangzhou, China, March 2008.